Table of Contents

- Introduction — 05
- Pawn Structure. — 09
- Find a plan in the middlegame. — 15
- There is always a weak point in the position, even if it is not immediately apparent. — 23
- Moves you can find if you change your way of thinking. — 28
- Recognize critical positions. — 31
- Tactics should always be present. — 36
- Candidate moves. — 40
- Intermediate Moves and Unpinning. — 48
- Method of elimination in defense. — 52
- The method of looking for Checks and Captures. — 56

- **The opponent's castling and your pawn chain.**   61
- **The prophylaxis.**   65
- **How to avoid mistakes in the middlegame.**   70
- **How Not to Play with the Queen.**   75
- **Pawn moves that weaken the castling position.**   78
- **Types of center.**   88
- **Evaluate the position.**   99
- **Endgames**   105
- **Tactic, tactic, tactic and tactic.**   125
- **Solutions.**   134

INTRODUCTION

It's great to see you again here, in the second part of this book series titled **"How To Play and Win at Chess Like a Master."**

In the first book, my intention was to acquaint you with masterful games played in the past.
Understanding the plans behind each move made by those chess masters is highly beneficial in your learning process, as it allows you to see things that, as a beginner, are very difficult to observe.

In this book, I will focus on fundamental strategies that you need to understand to improve your game. You will grasp the plans you can develop in a given position and what to do when you feel there's nothing to do at a certain moment in your games.

By finishing this book, you'll be able to play with more judgment, analysis, and deeper understanding.
I also want to tell you that there are things that, contrary to regular study that helps you improve your chess, can worsen, delay, or diminish your skills.

Currently, we live in an era where generations have a need for everything they want to do to be done quickly and easily.
Why do I tell you this?

Thanks to the internet, it's now too easy to play a game with anyone in the world. It's wonderful.
But it's very dangerous because most people resort to Blitz games.
Don't get me wrong, I'm not against Blitz games, but what I am against is a player who intends to improve their game, in the process, playing too many Blitz games, because having little time to think will affect your analytical ability and you won't be able to make correct moves, and your frustration will make you think you're not improving. If you lose many 3-minute, 2-minute, etc. games, don't worry, they're just Blitz.

To truly improve your chess level, you need to play slow games, analyze them, and study books that help you see what is difficult to see.

So, let's start studying the fundamental strategic topics you need to know in chess.

Best regards.

Alan H. Petrov

Pawn Structure

In this exercise, you will learn a lesson about one of the most important fundamentals in chess: pawn structure.

Set up the following position on your chessboard. Before you start reading, please take a few minutes to analyze the position on your own. Observe the board carefully, and once you have a clear idea, you may resume reading.
White to move.
Who do you think has the advantage?

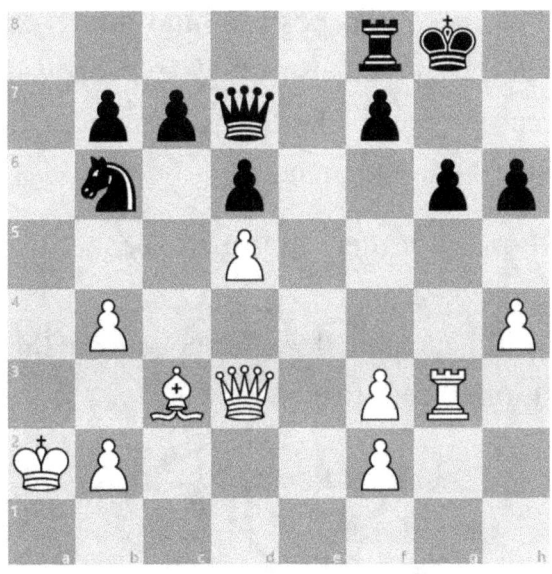

In this specific position, White has many open lines, allowing them to attack directly against the Black king. In fact, these open files and diagonals are what determine that White has the advantage in this position, leading to an immediate win.

White can play **1. Rxg6+**, checking the Black king and completely destroying their defense.

*(If Black captures the rook with **1...fxg6**, White continues with **2. Qxg6+** and Black can only cover with their queen using **2...Qg7**, and White delivers checkmate with **3. Qxg7#**.*
*Returning to White's rook sacrifice, if instead of capturing the rook, Black plays **1...Kh7**, White continues with **2. Rg7+** and after the Black king moves with **2...Kh8**, White checkmates with **3. Qh7#**).*

Let's go back to the initial position and analyze some details: White will win easily because they have open diagonals and files that directly attack the Black king's castling position, which is why they win the game.

But do you know why there are open files and diagonals? It's because of the pawn structure.

Now, what would happen if we removed all the pieces from this same position except for the pawns?

We can observe that the pawn structure is the same as before, with the same open diagonals and files still existing, but White no longer has the pieces to take advantage of them.

In this position, with no pieces to exploit those open columns and diagonals, Black is better, and White is destined to lose because their pawn structure is completely broken, with isolated and doubled pawns.
Let's see how the game could continue. White to move.

The game can continue with **1. Kb3, Kg7 2. Kc3, Kf6 3. Kd4, Kf5**
And Black's objective is to bring their king closer to the doubled pawns on the kingside to capture them, and then advance their f7 pawn to promote.
And if White tries to prevent the advance of the Black king with the move **4. Ke3**, Black would continue with **4...Ke5**, and then capture the white d5 pawn, winning easily.

In this example, White loses despite having open files and diagonals, which are of no use because they don't have pieces to take advantage of them.

Now, if in the starting position, we remove the major pieces and leave the minor pieces (the white bishop and the black knight), White would still be in trouble because they are going to lose the d5 pawn, which is being threatened by the knight, and the bishop is not able to protect it.

And Black would win very easily.

Let's go back to the starting position and leave only the pawns. That means we'll remove the black and white queens, the black and white rooks, and the black knight and the white bishop.

Leaving the position as follows:

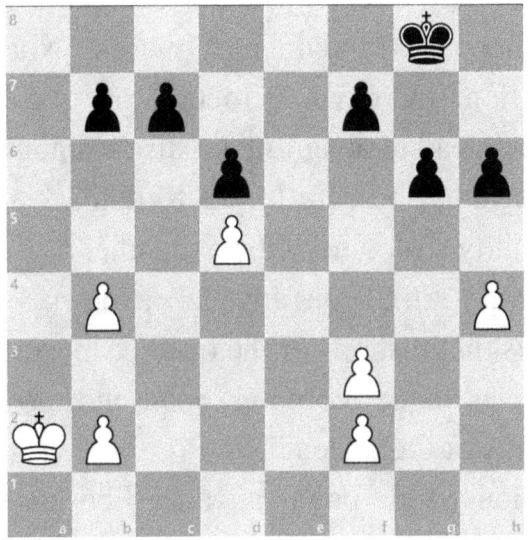

This position now on the board represents the pawn structure.

We refer to the arrangement of these pawns on the board as the pawn structure. It's important to note that the position of the pawns determines the power of the other pieces.

For example, in this position, White at some point advanced their pawn to d5, which allowed the opening of the long diagonal of black squares to be occupied by a bishop on c3.
Also, at some point in the game, they captured a piece on f3, thus opening the "g" file to be utilized by a rook.

However, with no pieces on the board, we can see that White is at a clear disadvantage.

Many novice players fail to grasp the importance of pawns. While they may not seem decisive at the beginning of the game, as pieces are exchanged and the game approaches its end, pawns become fundamental pieces. Not having paid enough attention to pawn structure in the opening and middlegame can lead to a significant disadvantage.

Therefore, you must be careful not to damage it too much to maintain a healthy structure in the endgame.

Remember, every time you advance a pawn, you open and close diagonals that bishops or the queen can exploit, and every time a capture is made, columns open and close that can be utilized by rooks or the queen.

Be very cautious with pawn advances because they cannot move backward.

Find a plan in the middlegame.

Let's analyze a position, and you'll understand how to formulate a game plan and how to determine the direction for your attack.

Place the following position on your physical chessboard or on your computer, analyze it for a moment, and try to find the next move for White.

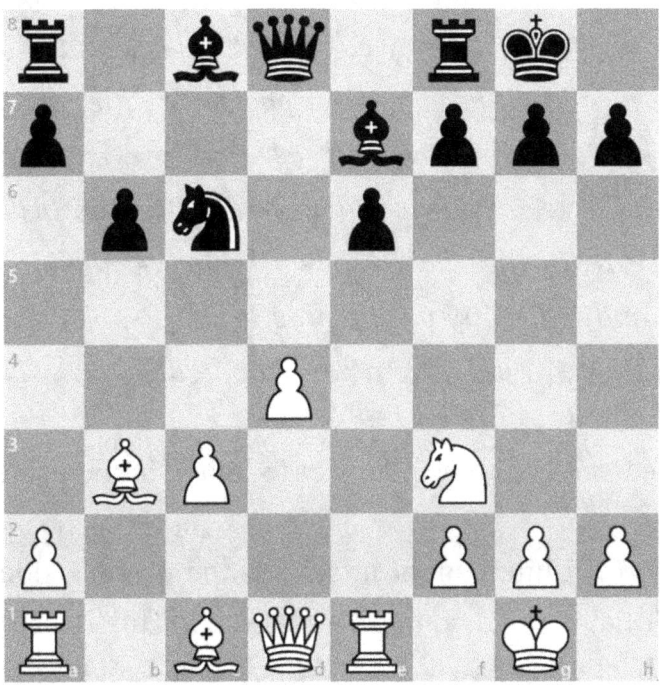

A fundamental rule of chess is that you must seize the initiative in the area of the board where you have more space or more pieces that can attack.

That being said, it's important to note that you should attack where you have some kind of advantage. In this position, White has more space in the center, as they have a pawn on the fourth rank, while Black's pawns only reach the third.

However, even though White has an advantage in the center, an attack there wouldn't be successful because Black has enough pieces to defend the center, making it very difficult for White to gain an advantage. For example:

*(if the game continues with **1. d5**, Black could respond with **1...exd5**, and after **2. Bxd5**, Black could continue with **2...Bb7**, developing their bishop while defending the knight. White would be left with a very weak pawn on c3, and Black could later pressure it with their rooks on the "c" file, also bringing their knight to a5 to control c4 and prevent White from advancing the weak pawn. Additionally, Black's bishop could go to f6, pressuring the c3 pawn while occupying the long diagonal and indirectly threatening White's rook on a1. In this scenario, Black would have a very good game).*

For this reason, White decides not to attack in the center now.

Furthermore, White cannot attack on the queen's side because they don't have an advantage in that area. This is due to White having a backward pawn on c3, which is a weakness that Black can exploit by pressuring the "c" file with their rooks and placing their knight on a5 to prevent White from getting rid of that weak pawn by advancing it.

Therefore, it must be said that White has an advantage on the kingside, as they have more space, and moreover, White pieces are pointing towards the kingside. Therefore, White will be able to mobilize all their pieces and achieve numerical superiority, ensuring a successful attack.

White has good control over the e5 square, preventing Black from advancing their pawn to this square and freeing their bishop on c8, which is currently a bad bishop and cannot defend the king's castling.

The Black knight on c6 also lacks good squares to move to the kingside, so two of Black's pieces will not be able to defend the kingside, and White will gain numerical superiority in this area.

With all that said, the correct move for White is **1. h4**, preparing the attack on the kingside.

*(If Black captures the pawn with **1...Bxh4**, White would continue with **2. Nxh4**, and after Black recaptures with **2...Qxh4**, White would continue with **3. Re3**, and then bring the rook to h3, from where it targets Black's kingside.*

Then White can quickly join the attack by bringing the white bishop to c2, together with the rook pressuring the h7 square.

*If Black continues with **3... Bb7**, developing their bishop, White can proceed with **4. Rh3**, targeting the Black queen. After Black retreats it with **4...Qf6**, White can continue with **5. Bc2**,* threatening to capture the h7 pawn. It wouldn't be possible for Black to protect their pawn by advancing with **5...h6** because now the h6 square is a sacrifice point for a White piece.

Therefore, White can swiftly continue with **6. Bxh6**, shattering the kingside defense, and Black would be in trouble. After capturing the bishop with **6...gxh6**, White can proceed with **7. Rg3+**, forcing Black to move their king with **7...Rh8**.

White would then continue with **8. Qd3**, threatening checkmate on h7. Black would have to defend with **8...Qf5**, but White would respond with **9. Qe3**, threatening to capture on h6 with the queen, delivering checkmate, and also threatening to capture the Black queen with the bishop. The only move for Black would be **9...Qf6**, retreating their queen and protecting the h6 pawn.

However, White would then continue with **10. Qe4**, renewing the threat of checkmate on h7. After Black covers the threat again with **10...Qf5**, White would proceed with **11. Qh4**, threatening to capture on h6 again, leading to checkmate.

Now, the Black queen cannot cover the h6 square as before because the f6 square is controlled by the White queen. Therefore, White would win. After **11...Qxc2**, White would play **12. Qxh6+**, and after Black covers the check with **12...Qh7**, White would continue with **13. Qf6+**, and if Black plays **13...Qg7**, White would deliver checkmate with **14. Qxg7#**).

(Let's go back a few moves in this variation, and instead of Black defending their h7 pawn by advancing it to h6, they play **5...g6**. In this case, White could continue with **6. Qg4**, gradually bringing pieces to the kingside, while two Black minor pieces are on the queenside and won't arrive in time to defend their king.

White threatens to drive away the Black queen with the bishop to g5, and the attack would be very strong.

Now, let's go even further back to analyze another variation. After White plays **1. h4**, Black responds with **1...h6**, trying to control the g5 square to prevent the White knight from going there. Now, White could continue with **2. Qd3**, continuing the attack, then follow up with Bc2, threatening Qh7# checkmate while also attacking the poorly defended h6 pawn. There's still the possibility of sacrificing the c1 bishop for that pawn, breaking through Black's kingside and making a decisive attack).

That's why in the game, after the move **1. h4**, Black responded with **1...Bb7**, developing their bishop and preparing the development of queenside pieces.

White continued with **2. Ng5**, already threatening to bring their queen to h5 for the attack and also placing their bishop on c2 to exert more pressure on h7.

(If now Black were to play **2...Bxg5**, removing the White knight, White could continue with **3. Bxg5**. In reality, the exchange of these pieces has benefited White because Black voluntarily decided to give up a good defender of their kingside. Now, White threatens the Black queen with their bishop, and if the Black queen retreats to another square, White can continue with **4. Qg4**. White would then be accumulating more pieces on the kingside and could easily move their rook. Additionally, White threatens to bring their bishop to f6, and the g-pawn cannot capture because it's pinned. This would give White a decisive advantage).

In the game, after the move **2. Ng5**, Black played **2...Na5**, attempting to eliminate one of White's powerful bishops and trying to create counterplay on the queenside, where they have an advantage.

The game continued with **3. Bc2**, preventing Black from capturing this valuable bishop while also placing it on a great diagonal aiming at the Black king.

Now, Black analyzed that they didn't want to further compromise their position by advancing their pawns, as it would weaken their kingside. Therefore, they exchanged their bishop for the knight with **3... Bxg5**, and White recaptured with **4. hxg5**. Now, Black played **4...Be5**, controlling the c4 square to prevent the White pawn from advancing, and forcing White to maintain that weakness. Black intends to bring their rook to c8 to pressure this weak pawn.

However, White continued with **5. Qd3**, threatening checkmate on h7, so Black responded with **5...g6**. White then replied with **6. Qg3**, bringing their queen to the kingside.

Black continued with **6...Nc4**, placing their knight on a good and centralized square. White proceeded with **7. Bf4**, and Black responded with **7...Rc8**. White then played **8. Rad1**, and Black made **8...b5**, trying to gain space on the queenside and also opening a diagonal for their queen to attack the weak pawns in that area.

White continued their attack with **9. Qh4**, and their attack became stronger, threatening to bring their rook to h3.

Black responded with **9...f5**, attempting to close the kingside. Now, White played **10. gxf6**, capturing en passant and keeping the game open to attack with their pieces.

Black continued with **10...Qxf6**, bringing their queen closer to defense and attempting to exchange queens to weaken White's attack. White proceeded with **11. Bg5**, preventing the exchange and attacking the Black queen. After Black retreated it with **11...Qf7**, White continued with **12. Rd3**, bringing their rook to the third rank to join the attack. Black played **12...Rc7**, preparing their rook to defend on the seventh rank and reinforcing the pawn on h7.

White continued with **13. Rh3**, and Black responded with **13...Nd6**, trying to bring their knight to f5, where it could be a good defender of the kingside and drive away the White queen.

White then played **14. Bh6**, attacking the Black rook. After **14...Rfc8**, White continued with **15. Bf4**, pinning the Black knight.

*(And now **15...Rd7** wouldn't be possible, because White would continue with **16. Bxd6**, and after Black recaptures with **16...Rxd6**, White would proceed with **17. Bxg6** and win, as they threaten to capture the Black queen, and it wouldn't be possible to capture the bishop with **17...hxg6** because White would deliver checkmate with **18. Qh8#**.*
*If after move **16. Bxg6**, Black captures the bishop with the queen with **16...Qxg6**, White continues with **17. Rg3**, pinning the Black queen).*

In the game, after move **15. Bf4**, Black continued with **15...Nf5**, attempting to attack the White queen, but White captured the knight with **16. Bxf5**, and Black resigned.
Because the rook on c7 is attacked by the bishop, White will win material and stay ahead.

There is always a weak point in the position, even if it is not immediately apparent.

Place the following position on your physical or virtual board, analyze it for a few minutes, and think about the best continuation for White.

In any chess position, there is always a weak point, even if it may seem imperceptible to us.
And throughout this series of books, I will insist repeatedly that your attack should be directed at these weaknesses, so focus on finding your opponent's weak spot.

The previous position would be somewhat balanced if the black pawn on c6 were still on its initial square at c7. However, since the pawn has already moved to c6, it has weakened the d6 pawn and also weakened the squares b6 and a7 on the queenside. So, now you have discovered what the weak points of the black position are.

So the white player played **1. Qb4**, initiating the attack against black's weaknesses.

*(Now it wouldn't be possible to remove the pawn from the threat with **1...d5** because white would continue with **2. Qb6**, threatening mate on c7, so black would have to play **2...Rd7** to avoid it, but white would resume the threat with **3. Qa7**, threatening mate on b8, so white would win).*

In the game, after the move **1. Qb4**, black responded with **1...Qd5**, defending the d6 pawn, while also bringing the queen closer to try to defend this area. White continued with **2. b3**, intending to bring the rook from a to d1 to defend the d4 pawn, before advancing their other pawn to c4 to attack the black queen.

The black player followed up with **2...h5**, attempting to counterattack on the kingside. The white player continued with their plan by playing **3. Rad1**, first defending their d4 pawn before advancing their c4 pawn.

*(If black were to play **3...h4**, attacking the white bishop, white would respond with **4. c4**, targeting the black queen. If black retreats to **4...Qf5**, white would simply continue with **5. Bxd6**, gaining an advantage.*

*And if after **3. Rad1**, black plays **3...c5**, attempting to attack the white queen, white would continue with **4. dxc5**, exploiting the exposed position of the black queen with their rook.*
*After **4...Qxc5**, white can proceed with **5. Qd2**, attacking the weakened black d6 pawn with three pieces. If black continues with **5...d5** to defend the pawn, white would play **6. Be5**, securing their bishop and threatening the black rook on h8. White could then continue their attack with moves like Re3 followed by Rc3, gaining a significant advantage, as the black e6 bishop is poorly placed due to its blocked mobility by its own pawns).*

In the game, after **3. Rad1**, black played **3...b5**, trying to control the c4 square. However, this move creates new weaknesses on the queenside, prompting white to play **4. Qa3**, attacking the weakened a6 pawn, which became vulnerable after the b pawn's advance.

*(If now black were to play **4...Kb7**, to defend the threatened pawn, the black king would be poorly placed on the "b" file, and white could continue with **5. c4**, attacking the black queen. It wouldn't be possible for black to play **5...bxc4**, because white would respond with **6. bxc4**, opening the "b" file for white to bring a rook to attack the black king.*

*And if after the move **5. c4**, black chooses not to capture the pawn to avoid opening the file, but instead plays **5...Qf5**, retreating their queen, white can continue the attack with **6. d5**, attacking the black bishop, putting white in a winning position. If black captures the pawn with **6...cxd5**, white would proceed with **7. cxb5**, completely disrupting black's queenside and mounting a strong attack. After black plays **7...axb5**, white would continue with **8. Qa5**, threatening to capture the b5 pawn. If black tries to defend it with the move **8...d4**, white could continue with **9. Rc1**, threatening to bring their rook to the c7 square and winning. If black continues with **9...Rc8**, white would respond with **10. Bxd6**, capturing the e6 pawn and adding another piece to the attack, putting white in a winning position. And after the move **8. Qa5**, black plays **8...Bd7** to defend the b5 pawn, white can continue with **9. Bxd6**, taking advantage of the fact that the black rook on d8 no longer defends the d6 pawn. White would then threaten to continue the attack with moves like Qc7 and Re7, pinning the bishop on d7. They can also bring their bishop to c5, from where they can threaten to enter with the queen on a7 or b6, gaining a winning advantage).*

That's why in the game, after the move **4. Qa3**, black played **4...Kc7**, attempting to bring a rook to the "a" file to counterattack the white queen in case it captures the a6 pawn. White continued with **5. c4**, attacking the black queen.
Black responded with **5...bxc4**, and white did the same with **6. bxc4**. Black replied with **6...Qxc4**, capturing a pawn while defending the a6 pawn. White continued their attack with **7. d5**. *(Of course, it would be a bad move for black to play 7...cxd5, because white would play 8. Rc1, capturing the queen on the next move).*
So in the game, after the move **7. d5,** black continued with **7...Bxd5**, and white followed up with **8. Re7+**, further invading the opponent's position.
Black retreated their king with **8...Kc8**, and white responded with **9. Bxd6**. Black blocked the white rook with **9...Rd7**, and white replied with **10. Qb2**, threatening checkmate on b8 and also capturing the undefended rook on h8.
Black followed up with **10...Rxe7**, capturing the white rook and giving the d7 square to their king to escape from checkmate on b8. But white continued with **11. Qb8+**, and black resigned.
*(Because after **11...Kd7**, white would continue with **12. Qc7+**, and it wouldn't be possible for black to play **12...Ke8**, because it would lead to checkmate with **13. Qxe7#**.*
*So, if after the move **12. Qc7+**, black tries to escape with their king by playing **12...Ke6**, white would continue with*
***13. Qxe7+**, and after **13...Kf5**, white would proceed with*
***14. Qe5+**, and then capture the black rook on h8, ending up with a rook advantage and an easy win).*

Moves you can find if you change your way of thinking.

In every chess game, there are different ways to find the right moves. However, because each player has a unique way of finding them, sometimes it becomes challenging to make them because the mindset is so focused on doing what you already know. This is called mental conditioning, and it prevents us from seeing beyond what we already know.

For beginners, it's more difficult to find the best moves for the knight than for the rook, just as horizontal movements are harder to find than vertical ones. Similarly, backward moves are more challenging to see than forward ones.

Now, please set up the following position on your physical or virtual board, analyze it for a few minutes, and continue reading to understand what I'm trying to explain.

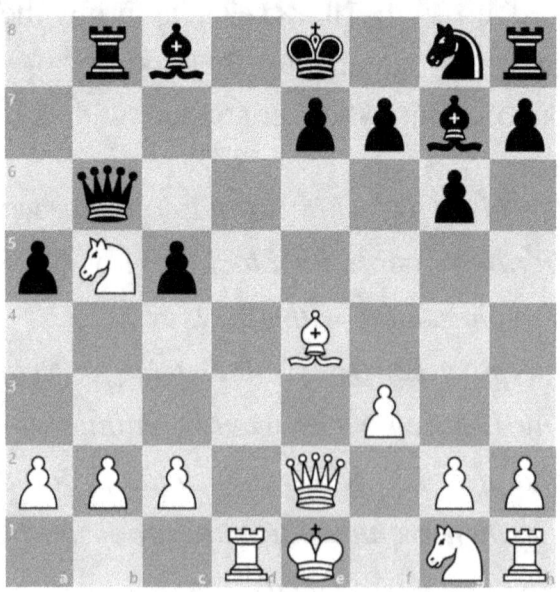

In this position, white played **1. Rd6**, and black resigned. Since white is threatening to capture the black queen, and the black queen has no squares to retreat to, black resigned for that reason.

(*It wouldn't serve to capture the white rook with **1...exd6**, because white would continue with **2. Bc6+**, delivering a double check with the bishop and the queen, and regardless of where the black king goes, white would deliver checkmate with **3. Qe8#**.*

*And if instead of capturing the rook with the pawn, black plays **1...Qxb5**, capturing the white knight, white would then continue with **2. Bc6+**, checking the black king, and in the next move, they would capture the black queen, thus gaining a winning advantage*).

However, after the move **1. Rd6**, both players forgot about a move.

(*The best move for black was **1...Bc3+**, checking the white king, and after white captures with **2. bxc3**, black could continue with **2...exd6**, capturing the white rook. Then, after white plays **3. Bc6+** with a double check from the bishop and the queen, black could continue with **3...Kf8**. Now, compared to the previous variation, black can escape checkmate with their king to the g7 square since they sacrificed their bishop, and it's black who would have the advantage. If the game were to continue with **4. Qe8+**, black would hide their king with **4...Kg7**.*

Now black would have a rook advantage for a bishop, and when they develop their knight, the white queen would be chased away, and they would activate their rooks to launch a quick attack).

This position serves as an example to the argument at the beginning, that due to each player having a peculiar way of dealing with threats, sometimes these same beliefs make the threatened player think that there is no solution to their position.

Most beginner players have the idea that threats are dealt with by capturing the threatening piece or simply having escape squares. When neither of these is possible, one must resort to intermediate moves, just as in this example. If black had made an intermediate move, checking the white king with their bishop while simultaneously freeing an escape square for their own king and gaining time to capture the rook, everything would have been resolved in time. However, lacking clear, easy-to-see, or direct moves, they decided to resign, when the game could have ended differently and very favorably for black.

Recognize critical positions.

There's a point in every chess game where you must carefully consider the next move. Recognizing when the position is at a critical juncture is perhaps one of the most challenging aspects of chess. Feeling the crucial moment when it's necessary to think deeply, as the rest of the game will depend on a single move.

To determine if the position you're in is critical, you must answer the following questions: Do you need to decide on a piece exchange? Is it necessary to decide on a change in the central pawn structure? What's the position after a series of forced moves?

After answering these questions, it's necessary to think about the next move, to formulate a game plan and make the best decision in the game.

Knowing when you're in a critical position is an advantage for your plans. Furthermore, after the game, you can analyze from that point and significantly improve your game by studying your mistakes or your opponent's mistakes.

Now, for this example, we'll analyze the following position, so set it up on your board and analyze it before continuing with the analysis, and think about how white should proceed.

Black's last move was **1...Re8**, which is a mistake because the black knight on f6 has no retreat squares. Simultaneously, there could be an attack against the poorly placed black queen.

If white wanted to capture the black queen, the white pawn on g2 should be on g4, the white knight should move to f3, and the pawn on h3 should be defended in order to capture the black queen.

*(Thinking about this, now the best move for white would be to play **2. Qf1**, preparing to execute the aforementioned plan.*

*If black tries to react with **2...h6**, intending to move their knight to h7 and clear the path for their queen to retreat, white would continue with their plan by playing **3. h3**, attacking the black queen. When the black queen retreats with **3...Qh5**, white would play **4. g4**, attacking the black queen again. After black moves the queen to **4...Qh4**, white would play **5. Nf3**, trapping the black queen).*

In this position, after black's move **1...Re8**, white continued with **2. g4**, *(with the idea that if black captures the pawn with **2...Qxh3**, then white would continue with **3. Rd3**, attacking the black queen, and after the queen retreats with **3...Qh4**, white would follow up with **4. Kg2**, intending to bring the other rook to h1 to trap the black queen, winning for white).*

So, after the move **2. g4**, black could have advanced the h pawn to h6, to then move their knight there and give the diagonal to their black queen to escape.

But instead of moving their pawn, black played **2...Nc6**, bringing their knight to the center, as it was in a position where its potential couldn't be utilized.
Black wants to eliminate the white knight on e5 and avoid threats. Now white played **3. Kg2**, defending their h3 pawn and threatening to move their knight to f3, winning the black queen. That's why black played **3...Nxe5**, capturing the white knight, and white recaptured with **4. dxe5**, trapping the black knight as it has no retreat squares, and white will win a piece.

Now black, knowing their f6 knight is lost, played **4...Nh5** to at least damage white's pawn structure. White captured with **5. gxh5**, and black played **5...Rxc4**, capturing a pawn in an attempt to compensate for the missing piece.

White continued with **6. Qf3**, threatening to invade with their rook to d7 to capture the f7 pawn, putting them in a winning position.

Black continued with **6...Rf8** to defend the f7 pawn. White played **7. h6**, with the idea of capturing the g7 pawn to disrupt black's kingside castle.

*(And it wouldn't be a good move for black to continue with **7...gxh6**, because white would then play **8. Kh2**, and white could occupy the open "g" file with their rooks, leading to a decisive attack).*

Therefore, instead of capturing the h6 pawn, black played **7...f5** to avoid white's attack and to try to advance their pawns and create counterplay.

But white continued with **8. Qg3**, threatening to capture on g7, which would be checkmate, and proposing a queen trade that would be beneficial for their position.

Black exchanged queens with **8...Qxg3+**, and white continued with **9. fxg3**. Black then brought their rook to the "c" file with **9...Rfc8**, and white captured the g7 pawn with **10. hxg7**.

Black followed with **10...Rc2+**, checking the white king.

White simply retreated it with **11. Kf3**, and black checked again with **11...R2c3**. White covered the check with **12. Be3**, and black captured another pawn with **12...Rxa3**. Black now has hope with their two pawns, which have a clear path and support each other.

But now white continued with **13. Rd7**, invading with their rook on the seventh rank. Black responded with **13...Rc4**, and white moved their king to free their bishop and be able to move it with the move **14. Kf2**.

And black continued with **14...Rc2+**, checking the white king, but white responded with **15. Re2**, covering the check and proposing a rook exchange. Black then played **15...Raa2**, and white exchanged with **16. Rxc2**. After **16...Rxc2+**, white replied with **17. Kf3**, and they have succeeded in exchanging pieces, simplifying the position, and will win with their piece advantage.

Now the a7 pawn is under attack, so black continued with **17...a5**, trying to advance their pawns quickly to create counterplay. But white continued with **18. Bh6**, defending the g7 pawn, and white threatens to check with their rook on the last rank, and after the black king moves, they can promote their pawn.

Black is forced to play **18...Rc8**, bringing back their rook to defend against that plan, but white continued with **19. Ra7**, and black resigned.

Since the black queen-side pawns cannot advance, and the black rook and king cannot move from their position, white would easily capture the black pawns, winning without complications.

Tactics should always be present.

We talk about tactics to refer to a sequence of moves aimed at gaining a specific advantage in a few moves, usually a material advantage or delivering checkmate. As Tartakower, one of the greatest chess players of his time, said, tactics is **knowing what to do when there is something to do, while strategy is knowing what to do when there is nothing to do.**
Tactics are present in every move but need to be discovered.

To find tactical resources, it's important to analyze positions in search of any forced combination that could give you a material or spatial advantage. Tactics will vary depending on the focus presented in the game, as it could be for a double attack, pawn break, blockade, attraction, discovered attack, passed pawn, X-ray attack, interception, deflection, pinning, overloading, defense annihilation, perpetual check, zwischenzug, and space clearance.

In many cases, a combination seeks to change a spatial advantage into a material advantage, so it's important to analyze your position for any imbalances that could help you gain an advantage or reduce your opponent's advantage.
First, you should look for your opponent's attack objectives, and once recognized, the next thing you should do is

find the necessary means to achieve those objectives.

Now, analyze the following position and place it on your physical or virtual board, take a few minutes to analyze it, and come back when you have an idea of what's happening on the board. **White to move.**

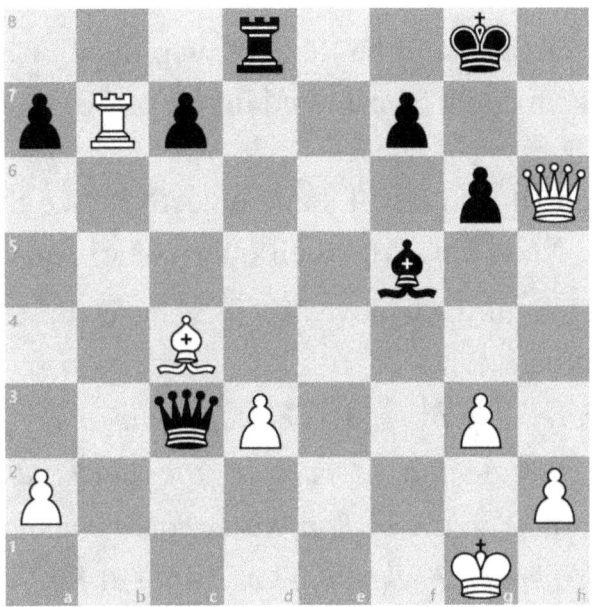

Let's do a material count. There are 4 black pawns and 4 white pawns, both sides have a light-square bishop, one queen each, and one rook as well. So, we can conclude that they are equal in material. Both sides have equal space, and apparently, the temporary position is balanced.

White's objectives could be the f7 pawn, the c7 pawn, and the a7 pawn, which could be captured in one or two moves.

Let's do a material count. There are 4 black pawns and 4 white pawns, both sides have a light-square bishop, one queen each, and one rook as well. So, we can conclude that they are equal in material. Both sides have equal space, and apparently, the temporary position is balanced.

White's objectives could be the f7 pawn, the c7 pawn, and the a7 pawn, which could be captured in one or two moves.

And black could also target some objectives like the d3 pawn, or attack the white king with their queen on e1 square.
But it's white's turn to move, so they have the advantage of taking the initiative.
*(If white were to play **1. Rxc7**, capturing one of the target pawns and threatening to capture the black queen with a discovered attack by the bishop, black could counterattack with **1...Qe1+**, checking the white king and after **2. Kg2**, black would continue with **2...Qe2+** and after **3. Kg1** the game would end in a draw due to perpetual check).*

So, for that reason, white played **1. Bxf7+**, sacrificing their bishop to gain time to capture their targets. The game continued with **1...Kxf7**, which is the only move.
Try to find the best move for white now...
Take 1 minute to think about it and continue:

White played **2. Rxc7+!**, checking the black king and simultaneously attacking the black queen on c3.

Now black's forced move would be **2...Qxc7**. Then white would play **3. Qh7+**, checking the black king, and when the black king moves, white will capture the black queen and gain a decisive material advantage.

The game continued with **3...Ke6**, and white, of course, played **4. Qxc7**, winning without any issues.

Black then played **4...Rxd3**, and after white played **5. Qxa7**, black tried to resist with **5...Rd1+**, checking the white king. After white played **6. Kf2**, black continued with **6...Rd2+**, giving another check to the white king. After white moved **7. Kf3**, black checked the king again with **7...Rd3+**, perhaps trying to achieve a draw by perpetual check.

But white played **8. Kf4**, and now the black rook can no longer give checks to the king because the d4 square is protected by the white queen on a7.

Black continued with **8...Kf6**, threatening to push their pawn to g5, which would be checkmate.

But White simply played **9. Qa6+**, checking the Black king and forcing it to retreat, and Black resigned.

(*Because after a move like* **9...Kf7**, *White could continue with* **10. Qc4+**, *and after* **10...Kf6**, *resuming the checkmate threat with the pawn, White can play* **11. h4**, *controlling the advance of the black pawn to g5. After this, White will advance their "a" pawn to promotion, and Black can do nothing to stop it*).

Candidate moves.

In this position, you will learn about plans, candidate moves, and how to calculate variations. Training your ability to calculate variations is fundamental to improving your chess game. If a player's analytical ability is inferior to their opponent's, they can lose a superior position in just a few moves.

The result of each game largely depends on the players' ability to calculate variations. To start training in analyzing variations, the first skill you need to focus on is OBSERVING THE POSITION. By doing this, you can understand its essence and find the strengths and weaknesses of BOTH SIDES.

Once you have observed and identified the weaknesses of each side, the next phase is to imagine what possible transformations in the position would be favorable. For example, piece exchanges, changes in pawn structure, etc. After this phase, you should think about specific moves that could achieve the transformations you envisioned as beneficial. These are commonly referred to as "candidate moves," as they are options you will consider and, at a certain point, lead to the advantageous position you imagined.

Analyze the following position, and familiarize yourself with it. Take a couple of minutes to think about what is happening on the board and start developing the three phases mentioned earlier.

Black to move.

To find the correct move in any analysis of any game, the first thing you must do is observe the position.

In this specific position, you can see that White is exerting pressure along two diagonals with their bishops, threatening to capture the Black rook on e3, and also threatening to capture the h6 pawn with their rook.

On the other hand, Black is threatening the g2 pawn from the long white diagonal occupied by their queen and bishop, and they have control of the open e-file.

Alright, we have now concluded the first phase, which is observation. Let's continue with the second phase, which is imagining. What transformations can occur in the position?

At the moment, no transformations in the pawn structure can occur, and the only potential piece exchange is Black's rook for the bishop on d3.

In such tense positions, the best and most recommended approach is to look for the most active moves.

Therefore, the most active moves for Black could be moving the rook to e2, invading the seventh rank, or moving the rook to e1, giving a check to the White king.

So, these would be the most promising moves for Black, or rather, the candidate moves for Black.

When you have two or three candidate moves, it is time to start thinking about variations for each of them.

To facilitate the calculation of variations, you can begin by considering the most forcing moves.

Forcing moves can be prioritized as follows:

1.- Checks.
2.- Captures.
3.- Threats.

We will start by analyzing variations with checks.

Variation #1.

We continue the game with **1...Re1+** giving a check to the White king, to which White would respond with **2. Rxe1**, and after **2...Rxe1+** White would continue with **3. Kf2**. Here, the Black rook would be attacked, and it is difficult to find a continuation for Black.

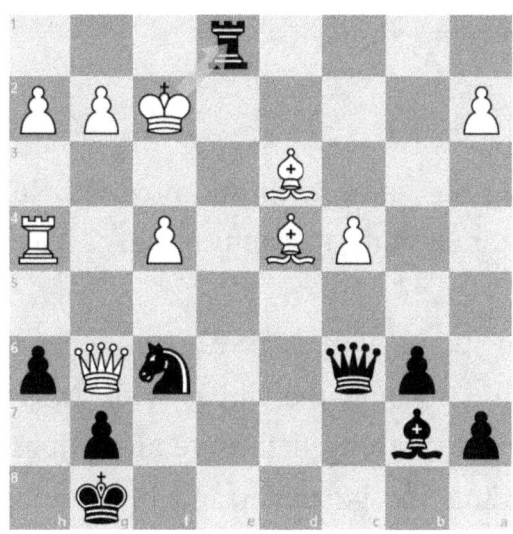
Example of Variation #1

If in this variation, Black continues with **3...Rd1** attacking the bishop on d3, White can respond with **4. Ke2**, which would defend their bishop and also attack the black rook again. Alternatively, after Black plays **3...Rd1**, White can directly play **4. Bxf6** and would be winning, as they are threatening mate on g7 and the black pieces are defenseless.

It wouldn't be possible for Black to play **4...Qxf6** because White would then continue with **5. Qe8+** and the black king cannot escape because the bishop on d3 attacks its only escape square. Therefore, Black would have to cover the check with **5...Qf8**, but then White would respond with **6. Bh7+** and the black king must move away from its queen, allowing White to capture it on the next move. So, after analyzing this variation with a rook check, we conclude that it's more beneficial for the opponent than for us.

Let's now analyze the threat.

Variation #2.
Black continues the game with the move **1...Re2**, entering the seventh rank, threatening to capture the g2 pawn and win the white queen.

White would respond with **2. Bxe2**, removing the black rook, but Black would resume the threat with **2...Rxe2**. However, White would have a strong defensive and offensive move with **3. Rg4**, defending the g2 pawn and threatening mate on g7. Therefore, White, who now has a rook and a pawn advantage, has completely halted Black's threat and it's White who is strongly attacking in the position.

Example of Variation #2

Now, let's analyze variation #3, which is the capture.

Variation #3.

Black continues the game with **1...Rd3,** and White would recapture with **2. Rxd3**. In this position, we can notice that, compared to variation #1, the white light-square bishop is not on the board, so Black could continue with **2...Re1+** giving check to the white king, and after **3. Kf2**, Black can proceed with **3...Ne4+** giving check to the white king, and Black would capture the white queen on the next move.

Example of Variation #3

The game continued with **4. Kxe1**, and Black captured the white queen with **4...Qxg6**, giving Black a decisive material advantage. Additionally, the white king returned to the center of the board, and the white pieces are very poorly coordinated.

White continued with **5. h3**, intending to bring their rook to g4 and attack the black queen, but Black responded with **5...Nf6**, retreating their knight to defend the g4 square and simultaneously attacking the exposed white rook on d3.

White continued with **6. Rd2** to remove their rook from the threat, and Black played **6...Qg3+**, checking the white king and threatening to capture the rook on h4.

White defended with **7. Bf2**, and Black simply continued with **7...Qxg2**, capturing a pawn, and Black is winning.

White responded with **8. f5**, attempting to activate their rook on h4; however, Black played **8...Qh1+**, checking the white king, and after **9. Ke2**, Black checked again with **9...Qf3+**, forcing White to retreat their king with **10. Ke1**.

With this maneuver, Black improved the position of their queen by removing it from the rank where the white rook could attack it with a discovered attack. Now, Black played **10...Ne4**, centralizing their knight and preventing the rook on h4 from coordinating with the other white rook. Now, Black is threatening to capture the d2 rook.

(It would be a bad move for White to play **11. Rc2**, *maintaining the defense of the f2 bishop, because Black would continue with* **11...Qh1+** *and after* **12. Ke2**, *Black would capture the bishop with* **12...Nxf2**, *and after* **13. Kxf2**, *Black would continue with* **13...Qg2+**, *and in the next move, Black would capture the white rook, giving Black a significant advantage).*

Therefore, in the game, after the move **10...Ne4**, White played **11. Rg4**, but Black continued with **11...Nxd2**, and White finally resigned because Black's material advantage is superior, and Black would win without difficulty.

Intermediate Moves and Unpinning

Now, let's analyze a complete game, in which you'll learn the technique of deflection and the importance of making intermediate moves.

The game started with moves **1. e4, e5 2. Nf3, Nc6 3. Bb5**. The Spanish Opening.

Black responded with **3...a6**, attacking the white bishop, intending to drive it back. White chose the exchange variation with **4. Bxc6**, and Black captured with **4...dxc6**, recapturing with the d-pawn to keep their queen-side pawns united and simultaneously allowing the quick development of their queen's bishop.

*(Now it wouldn't be a good move for White to capture the unprotected pawn with **5. Nxe5** because Black would continue with **5...Qd4**, attacking the knight on e5 and the pawn on e4, and Black would regain material and gain an advantage.*
*If White were to retreat the knight with **6. Nf3**, Black would continue with **6...Qxe4+**, capturing the central pawn of White's, and the center would be open with Black having the bishop pair, and you must remember that open positions benefit the player with the bishop pair).*

In the game, after the move **4...dxc6**, White played **5. Nc3**, developing their queen's knight, so now White can capture the e5 pawn. Black defended it with **5...f6**.

Then, White played **6. d3**, to clear the path for their c1 bishop. White aims to keep the game closed since they have lost the bishop pair and need to prevent the opponent's bishops from becoming active along open diagonals. Thus, they will try to strengthen their knights in a closed position.

Black continued with **6...Bg4**, developing their queen's bishop and pinning White's f3 knight. White responded with **7. h3**, attempting to break the pin. Black withdrew their bishop with **7...Bh5** to maintain the pin. Now, White played **8. Be3**, completing the development of their minor pieces. White can then follow up with Qd2 to prepare queenside castling and advance their g2 pawn to g4 to break the pin.

Black continued with **8...Qd7**, but this move is a mistake since their h5 bishop is unprotected. White can capitalize on this situation to unpin their knight. Now, White played **9. Nxe5**, capturing a central pawn of Black's and attacking the opposing queen. This way, White freed themselves from the pin and is attacking the unprotected h5 bishop.

*(And it wouldn't be a good move for Black to continue with **9...fxe5**, capturing the white knight, because White would then play **10. Qxh5+** and then capture the e5 pawn, giving White a significant advantage).*

In the game, after the move **9. Nxe5**, Black responded with **9...Bxd1**, capturing the white queen, as their bishop had no retreat squares. White then continued with **10. Nxd7**, capturing the black queen, and now both of Black's bishops are under attack, giving White the advantage.

*(If Black were to play **10...Kxd7**, White would capture the bishop with **11. Rxd1** and would remain with a pawn advantage, with more developed pieces and more space on the board, so we can say that White would have a clear advantage).*

That's why in the game, after move **10. Nxd7**, Black responded with **10...Bxc2**, capturing a pawn to try to maintain material balance, but White replied with **11. Nxf8** capturing the bishop and would win material.

*(And it wouldn't be a good move for Black to continue with **11...Bxd3**, because White would simply proceed with **12. Ne6**, withdrawing their knight, and White would be left with a piece advantage).*

So in the game, after move **11. Nxf8**, Black captured with **11...Kxf8**, and it seemed like Black had regained material and threatened to capture the d3 pawn.
But White played move **12. Kd2** and Black resigned.
Since the white king defends the d3 pawn and simultaneously attacks the opponent's bishop, which has no retreat squares, White would be left with a piece advantage and win the game.

This game serves as a perfect example to explain how a pin can change the situation of the position for the player who does not know how to use it correctly. You must always be careful to ensure that all your pieces are protected by others to avoid becoming a victim of discovered attacks and losing material.

Method of elimination in defense.

In the following position, you will learn about the method of elimination in defense.

Set up the following position on your board and analyze it for a few minutes, then think about the next move for Black.

The position of the Black pieces is quite delicate as they are down a pawn, and the White pieces have good control of the position.

White could increase the pressure by moving their king to a safer square like g2 and continue by bringing their knight to d4, threatening to capture the c6 pawn. Black has several problems to solve if they don't want to end up losing the game

If you were the Black pieces... How would you defend against the threats?

...

Quickly looking at the position, any beginner player might think that Black is simply lost. However, in this position, there's a move for Black that might surprise you.

Black played **1...g5!**, which is an incredible move found after discarding all the alternatives Black could have made.

*(If Black had decided to continue the game with a more direct move like **1...c5**, attempting to capture the pawn on b4 to destabilize White's queenside pawns and trying to defend, although White could proceed with **2. Nxc5** attacking the Black queen, and after Black captures the knight with **2...Bxc5**, White could continue with **3. Qxc5**, and Black could try to achieve perpetual check to aim for a draw with **3...Qd1+**, since it wouldn't be possible for White to play **4. Kg2** because Black would then follow with **4...Nf4+** giving check to the White king, resulting in a draw by perpetual check. After White captures with **5. gxf4**, Black continues with **5...Qg4+**, and after **6. Kf1**, Black would play **6...Qd1+**, the White king retreats with **7. Kg2**, Black continues with **7...Qg4+**, and if the king moves to **8. Kh2**, Black captures with **8...Qxh4+**, and the White king cannot escape the checks, resulting in a draw.*

And if after the move **6...Qd1+** *White responds with* **7. Kh2**, *thus avoiding perpetual check and White would win).*

But in the position, after the move **1...g5!**

White replied with **2. hxg5**, capturing the Black pawn, but Black continued with **2...c5** and is threatening to capture the White pawn to free their position.

(Now it would be a bad move for White to play **3. Nxc5**, *because Black would then continue with* **3...Bxc5**, *capturing the White knight, and White cannot capture the bishop with the pawn because they would lose the queen, so they would have to capture with the queen with* **4. Qxc5**, *but Black would continue with* **4...Qd1+** *giving check to the White king, and Black would achieve a draw.*

And if White tries to avoid the perpetual check with **5. Kg2**, *Black would then continue with* **5...Nf4+**, *and the perpetual check cannot be avoided, since after White captures the knight with* **6. gxf4**, *Black continues with* **6...Qg4+**, *and after* **7. f1**, *it would follow* **7...Qd1+** *and so on).*

That's why in the game, after the move **2...c5**, White didn't capture the black pawn, but instead played **3. Kg2**, attempting to place their king on a safer square to avoid Black's perpetual check.

Black continued with **3...cxb4**, eliminating a white pawn, and after **4. axb4**, Black continued with **4...Qa2**, threatening to capture the bishop on b2. White then retreated that bishop with **5. Bc1**, and Black followed with **5...Bxb4**, eliminating all the pawns on the queenside with good chances of achieving a draw, since the remaining pawns are on the same flank.

As you can see, at the beginning of the position, Black had their pieces very passive, but as the moves progressed, they managed to coordinate and activate them, while White lost all the initiative they had.

White continued the game with **6. Qc8+**, checking the black king, and Black responded with **6...Kg7**. White centralized their knight with **7. Nd4** to try to attack the black king. However, now Black continued with **7...Bc3**, threatening to capture the white knight. With Black's pieces very active, they easily secure a draw.
Now White tried to unbalance the position with **8. Nxe6+**, attacking desperately; however, White no longer has any chance of winning the game.
Black responded with **8...fxe6**.

(And now it would be a very bad move for White to play
9. Qxe6, capturing the pawn, because Black would continue with 9...Ne3+, giving check to the white king, and in the next move, they would capture the black queen due to the discovered attack).

After the move **8...fxe6**, White continued with **9. Qd7+**, and Black followed with **9...Kh8**. Then, White played **10. Qe8+**, giving another check to the black king, and finally, both players agreed to a draw.

The method of looking for Checks and Captures.

Analyze the following position for a few minutes and come back when you have the best moves.

How would you continue with White?

There's a method to find combinations and the best candidate moves which involves finding all possible checks, even if these checks aren't feasible. Strange, isn't it?
What I mean is that you can search for checks even by imagining that your pieces 'pass through' those in their way to the king.

For example, in the position you just saw, there are no possible checks with legal moves, right?
However, you have to imagine as if the white pieces could pass through and give checks.

Here are the examples:

1.- If the white queen could go to f8, it would be check.
2.- The rook on c2 would check with Rxc8+.
3.- The rook on e1 would check with Re8+.
4.- The bishop on b3 would check with Bxe6+.

Of course, these checks are not possible because there is no check with a legal move.

Now, we will do the same method but with captures. There are captures with possible legal moves and moves that we have to imagine that the pieces "pass through each other".

These are the captures with possible moves (only captures, even if the moves are bad):

1.- Capture the knight with the queen (Qxd5).
2.- Capture a pawn with the queen (Qxf5).
3.- Capture a pawn with the rook (Rxc6).
4.- Capture a pawn with the rook (Rxe6)
5.- Capture a pawn with the bishop on black squares (Bxh6).
6.- Capture a pawn with the bishop on white squares (Bxd5).

These are the captures with moves that are not possible but that we could make if our pieces "passed through" the pieces that interfere with our path:

1.- Capture the rook with the queen (Qxf6).
2.- Capture a pawn with the queen (Qxc6).
3.- Capture a rook with the rook (Rxc8).
4.- Capture a pawn with the bishop with (Bxe6+).

Once you've located all the captures and checks (including captures and checks "passing through" pieces), you need to start analyzing each move, beginning with the most forcing ones. That is, those that most compel and limit the opponent to decide among fewer options or just one, simplifying the calculation process because the number of variations is greatly reduced, as most of the opponent's responses will be nearly forced.

In order, we would start by analyzing the checks, but since there are no direct checks in this example, we'll move on to the captures.

Let's start with the most forcing capture, **1. Qxd5**, since sacrificing the queen would almost compel the opponent to capture it. In this example, the opponent could capture it with either of their two pawns protecting the d5 square. However, the c6 pawn is pinned since the c8 rook is undefended. Yet, as the queen is the most powerful piece, even if we manage to capture the c8 rook, we cannot see a variation where we might gain an advantage afterward. Therefore, the most forcing capture would be discarded.

Let's then move on to the second most forcing move.

The move would be **1. Bxd5**, as capturing the knight with the bishop would compel a forced response from the opponent to restore material balance.

In contrast to capturing the knight with the queen, now that the knight has been captured with a bishop, the black side cannot freely choose any pawn for recapture. If they capture with the c6 pawn, the black rook on c8 would be captured without issues, and we would win a rook. Therefore, it is compulsory for black to capture with the e6 pawn, with **1...exd5**. However, after this capture by black, it becomes challenging for white to find a good continuation with the attack.

But now, the idea arises that if either of the two pawns defending the knight were to disappear, it could complicate the black position.

For example, if the e6 pawn were absent and white captured the knight with **1. Bxd5**, if black wanted to reclaim the material by capturing with their c6 pawn, then they would lose the c8 rook, and white would win easily.

With this in mind, we can think about a way to capture the e6 pawn to eliminate a defender of the d5 knight and then capture it later.

So the next capture to consider would be **1. Rxe6**, and then we would need to consider what would happen after Black recaptures with **1...Rxe6?**

Now White could play **2. Bxd5**, pinning the black rook on e6, and Black couldn't recapture with **2...cxd5** either because they would be losing the rook on c8 to **3. Rxc8+**, regaining the rook and gaining an advantage.

And now you can see how easy it was to find the solution using the method of finding possible and impossible checks and possible and impossible captures. This method greatly clarifies your mind towards possibilities, and from these possibilities emerge combinations and tactics to better understand the position. This method of captures and checks is known as the Purdy method.

So the solution and the move played by White was **1. Rxe6!**, winning a pawn. Black responded with **1...Rxe6**, and White continued with **2. Bxd5**, pinning the rook on e6.

If Black wanted to resist, they should play **2...Re8** to defend the rook, but then White would play **3. Bxe6+**, capturing the rook. After Black recaptures with **3...Rxe6**, White could continue their attack with **4. d5**, advancing their isolated pawn while removing its weakness. It's important to remember that an isolated pawn, especially with many pieces still on the board, is a weakness.

Furthermore, with this move, White threatens the rook on e6 and the pawn on c6.

Black would have to play **4...cxd5**, and White succeeded in opening the "c" file for their rook, and they could then proceed with **5. Qxf5**, attacking the black rook and the pawn on d5, threatening to bring their rook to c8 with a very strong attack. Black resigned.

The opponent's castling and your pawn chain.

The following game serves as an example of how we can attack the opponent's castled position while maintaining our flexible pawn chain.

The game started with the moves:
1.d4, Nf6 2. c4, e6 3. Nc3, Bb4.

The Nimzo-Indian Defense.

The game continued with **4. Qc2,** preventing Black from doubling pawns on the "c" file and controlling the e4 square.

Black castled with **4...O-O,** and White continued their development with **5. Nf3**. Black played **5...d6**, and White developed their bishop with **6. Bd2**, releasing the pin on their knight. Black responded with **6...Nbd7**.
Now White drove away the bishop on b4 with **7. a3**, but Black captured with **7...Bxc3**, and after White's recapture with
8. Bxc3, Black played **8...Qe7**.
White continued with **9. e3**, making way for their bishop, and Black played **9...b6** to develop their bishop by fianchettoing. White responded with **10. Be2**, and Black played **10...Bb7**. White continued with **11. b4**, starting their attack on the queen's side.

Now... Take a few minutes and...
How would you proceed with Black?

In the position, White has initiated their attack on the queen's side, and Black must counterattack on the king's side. So their plan is to advance their king's side pawns so that when the king castles, Black's attack will already be underway.

Black continued with the move **11...Ne4**, centralizing their knight and allowing the advance of the f7 pawn to begin the attack on the king's side.

White responded with **12. Bb2**, relocating their bishop to maintain the bishop pair, and Black continued with their plan with **12...f5**. White castled kingside with **13. O-O**

Black continued preparing their attack with **13...Rf6**, intending to transfer the rook to the sixth rank to pressure the opponent's castled position.

White followed with **14. d5**, attacking the black rook indirectly and blocking the diagonal for the bishop on b7. Black responded with **14...Rh6**, and White replied with **15. dxe6.**

Analyze the position up to this point again and...
How would you proceed with the Black pieces?

Capturing the pawn with the queen isn't the best option.
The optimal move is to continue with **15...Nf8**, allowing for the capture with the knight. This way, the knight relocates to another square and can more easily join the attack. Additionally, the queen continues to control the dark squares diagonal, which can be utilized at some point.

White continued with **16. g3**, aiming to control the h4 square and prevent the black queen from going there, where it would threaten checkmate on h2.

Black captured the pawn with **16...Nxe6**, and White followed with **17. Nd4**, attempting to exchange the knights so that their bishop on d2 could move to defend the castled position.

Analyze the position and try to find the continuation for Black.

Actually, speaking in terms of strategy, White's move had a good intention, which is to activate their bishop on e2 and at least have it function as a good defender of their king's castled position.

However, White overlooked a tactical detail that Black will take advantage of.

Black continued with **17...Qh4**, threatening checkmate on h2.

*(It's not possible to capture the queen with **18. gxh4** because then black would proceed with **18...Rg6+** and after **19. Kh1**, black would end the game with **19...Nxf2#**, resulting in checkmate, as the king is in double check by the bishop and the knight).).*

So, after move **17...Qh4**, white defended with **18. Nf3** to protect h2, and black responded with **18...Qh3**. Then, white played **19. Rfd1** trying to provide some shelter for their king and to bring their bishop to f1 to drive away the black queen.

Once again, analyze the position and try to find the best continuation for Black.

Black continued with **19...N6g5**, bringing another knight into the attack.

With this move, black threatens to capture the knight on f3, which is the only piece guarding h3, preventing checkmate in a few moves. So, white should not allow this exchange and plays **20. Nh4**, attempting to block the h-file. However, black proceeded with **20...Rxh4**, and after **21. gxh4**, it's your turn again to analyze the position and find the continuation for black.

Black played the move **21...Ng3**, which unleashes the action of the bishop on b7 and threatens checkmate with the queen on g2, causing white to resign, as they saw no way to thwart the threats.

In this game, you've seen an example of how to attack the opponent's castling by advancing the right pawns to launch a forward attack when your opponent intends to castle.

Taking the time to analyze various positions within a single game helps expand your imagination by envisioning lines in later positions that you wouldn't have seen in earlier moves.

Take your time in each game; remember that with every move, the position changes.

The position isn't static; it's always evolving. Like any event in life, if it changes and you don't adapt to the new environment, it's very likely you won't survive.

The prophylaxis.

Prophylaxis involves preventing your opponent's intentions. It's more than just defense; it's synonymous with anticipating your rival's plans, even before they create threats.

In other words, it's about hindering their plans and thwarting their intentions before they materialize.

If you can play your games with this in mind, rest assured your level will increase rapidly, and your results will undoubtedly improve.

To get accustomed to thinking prophylactically in your games, you should constantly ask yourself the following:

What does my opponent intend?
What would they play if it were their turn?
Which move by my opponent will disrupt my position?

So, we can conclude that prophylaxis in chess is based on identifying the opponent's plan before it unfolds.

Let's better understand this concept with the following example:

The latest move by white was **1. Rf3,** with the idea of attacking on the king's side, and white would like to follow up with moves like **2. Rh3** and then **3. Qh5** to attack square h7.
Alternatively, they can play **2. h3** to break up black's pawn structure with move **3. g4.**

As for black, their plans involve expanding on the queen's side with moves like **1...a6** to drive away the bishop, **2...b5,** and continue with **3...a5** and **4...b4** to open columns on the queen's side and attack through this area.

Given the plans of both sides, consider the best continuation for black.

In this position, the plan for black could have started with the move to drive away the bishop with a6; however, black continued with **1...h5.**

With this move, black prevents the advance of the white pawn to g4, where white could have initiated a breakthrough. What black is doing is anticipating white's plans before starting their own attack on the queen's side.

This is prophylaxis: black is preventing white's attack on the king's side so that they can later attack freely on the queen's side.

(The white side couldn't play **2. Rh3** *immediately attacking the h5 pawn because black would defend it calmly with* **2...g6**, *and it would be a bad move for white to break with* **3. g4**, *since after black captures with* **3...hxg4**, *black would then have a pawn advantage, as well as dominate the opened "h" file).*

Therefore, after move **1...h5**, white played **2. Ref1**, and black continued with **2...Rh6.**

A curious move by black, as they don't place the rook on an open file but rather behind their own pawn, with the plan of preventing white from advancing their pawn to g4 because the rook is aiming at the h1 king like an X-ray.

(Now white couldn't play **3. h3** *and* **4. g4**, *because then black would exchange pawns, and the file would open for black's rooks, giving them an advantage).*

Using prophylaxis, black has thwarted white's plans on the king's side to then proceed with their own on the queen's side.

The game continued with **3. Be1, g6 4. Bh4, Kf7 5. Qe1**, and now black played **5...a6**.

With white's attack on the king's side neutralized, black now begins their own attack.

The game continued with **6. Ba4, b5 7. Bd1, Bc6**, and black aims to advance their pawns to a5 and b4. With the bishop on c6, they control square a4 to prevent the return of the white bishop, which could block the pawns.

The game continued with **8. Rh3**, and black pursued their plan with **8...a5**. White responded with **9. Bg5**, attacking the black rook, which black withdrew to **9...Rh8**.

White still struggles to break through on the king's side while black's attack on the queen's side remains potent.

White continued with **10. Qh4,** and black played **10...b4**, aiming to open lines on the queen's side and pave the way for their pieces.

White then played **11. Qe1**, and black responded with **11...Rb8**. White tries to redirect their pieces to the other flank with **12. Rhf3**, and black continued with **12...a4**.

The game proceeded with **13. R3f2, a3**.
*(Now white couldn't play **14. bxa3**, because black would respond with **14...b3**, gaining a dangerous passed pawn).*

Therefore, white played **14. b3**, but this move loses a pawn. Black followed with **14...cxb3**, and white recaptured with **15. Bxb3**. However, black played **15...Bb5**, attacking the rook, so white removed it with **16. Rg1**, and black continued with **16...Qxc3**.

White captured with **17. Qxc3**, and after **17...bxc3**, black has won a pawn and possesses a dangerous passed pawn.

White blocked the pawn with **18. Rc2**, and black responded with **18...Rhc8** to defend the pawn. White then continued with **19. Bh4**, and black countered with **19...Bd3**, attacking the rook that was blocking the pawn so it could advance.

White proceeded with **20. Rcc1**, and now black played **20...Rxb3**, and after white recaptured with **21. axb3**, black played **21...a2**, and white resigned.

White resigned because black threatens to bring their light-square bishop to b1 and then promote the "a" pawn.

Therefore, the only move for white would be **22. Ra1**, but then black could play **22...c2**, threatening to advance the "c" pawn and promote it to a queen, and there's no way to stop it.

For example, if white blocks this pawn with **23. Rgc1**, black would respond with **23...Ba3**, and black would win.

But if after move **22...c2**, white plays **23. h3**, black would play **23...c1=Q**, and after **24. Rgxc1** and **24...Rxc1+**, **25. Rxc1**, black would play **25...Bb1**, and in the next move, promote the "a" pawn, winning the game.

How to avoid mistakes in the middlegame.

It's normal in the middle game of a match to feel unsure about what to do, as the position appears calm, and you haven't quite decided on your plan for the remainder of the game, or you might think any waiting move is suitable.

In the following position, you'll understand that when you're uncertain, it's better to move a piece other than a pawn, as it's easier to make a mistake by moving a pawn. As you already know, pawns can't move backward, so you wouldn't be able to retract the move and correct the mistake.

Especially in time pressure situations, it's advisable to avoid moving pawns because advancing a pawn inevitably creates weaknesses.

Analyze the following position and grasp what's happening within it.

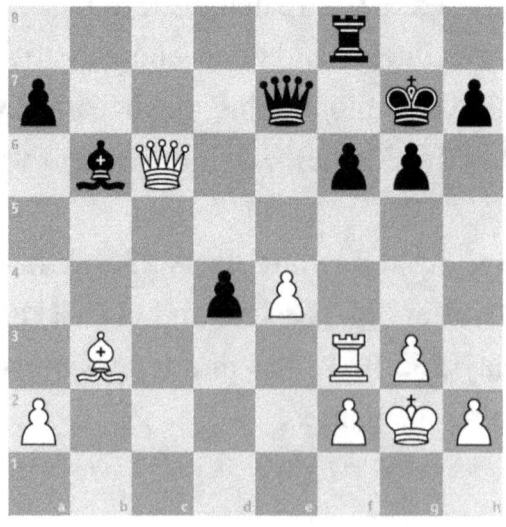

(In the position, Black could have continued with the move **1...Rd8***, placing their rook behind the passed pawn to support its advance, and the position would have been balanced).*

However, perhaps due to time pressure, Black attempted a move they believed to be more active. So in the game, Black played **1...h5**, advancing a pawn, but this move is a mistake because it weakens the king's flank, weakens the g6 square, and doesn't contribute anything positive to Black's position.

White continued with **2. h4**, preventing the black pawn from advancing further. Black then proceeded with **2...Rd8**, placing the rook behind the black pawn. White responded with **3. Bd5**, blocking the rook's action.

Black then played **3...Qd6**, suggesting a queen exchange, but White declined this exchange with **4. Qc2**. Now, with this move, White begins to target the g6 pawn through X-rays, exploiting the weakened pawn resulting from Black's advance to h5.

Black continued with **4...Rd7**, and then Black proceeded with **5. Qc8**, invading the last rank.
Black had no choice but to retreat the rook with **5...Rd8**, and White responded with **6. Qa6.**

(Perhaps it would have been best for Black to return the rook to **6...Rd7***, protecting the seventh rank, and the position would have been more or less balanced).*

It's true that Black has certain weaknesses in the position, but they aren't yet overly dangerous weaknesses.

However, after White's move **6. Qa6**, Black continued with **6...f5**, and the pawn cannot be captured because the white bishop on d5 would be captured.

But by advancing the pawn to f5, all Black achieved was further weakening the position of their king.
White continued with **7. Bc4**, withdrawing their bishop

*(It wouldn't be possible for Black to play **7...fxe4**, because White would then follow up with **8. Rf7+** and then bring their queen to b7, resulting in a decisive attack for White).*

Therefore, Black responded with **7...Qc6**, attempting to capture the e4 pawn with the queen. But White continued with **8. exf5**, and Black played **8...Rf8**, trying to exploit the pinned white rook. So, Black aims to capture the f5 pawn with the rook and then capture the white rook.
The whites played **9. Bd3,** defending the f5 pawn from being captured by the black rook. Now, the whites threaten to advance their "a" pawn to a5, attacking the pinned black bishop. If it retreats, the black queen would then be lost.

Black continued with **9...gxf5**, and White responded with

10. Bb1, intending to bring their queen to e2, to defend their rook and free their king to move to another square, thus freeing their rook.

Black continued with **10...Rf7**, defending the seventh rank, and White brought their queen to **11. Qe2**, now having the advantage.

White will proceed with Kh2 to unpin their rook, and then capture the weak pawns of Black on the kingside.

Black followed with **11...Bc7**, and White responded with

12. Qd2, threatening to capture the black d4 pawn and bring their queen to g5, increasingly invading the kingside.

Black played the move **12...f4**, and White captured the d4 pawn with **13. Qxd4+**, to which Black covered the check with **13...Rf6**. White responded with **14. Qd3**, threatening a strong entry with the queen to h7.

The whites are taking advantage of the weaknesses created because the blacks have advanced their pawns imprudently.

Now, the blacks played **14...Kf8**, anticipating the whites' plan, but the whites proceeded with **15. Qh7**, invading the kingside immediately.

The blacks followed with **15...Be5**, and the whites made the move **16. Qh8+**, leading the blacks to resign the game.

*(If Black responds with **16...Kf7**, White would continue with **17. Qxh5+**, threatening to capture the Black bishop. Therefore, the only move would be **17...Ke6**, but White could then play **18. g4**, gaining a two-pawn advantage and leaving the Black king very exposed. This would give White a significant advantage, as the pawns support each other and either one could potentially be promoted).*

*(If after **16. Qh8+**, Black plays **16...Ke7**, then White would go to the same square with **17. Qxh5**, threatening to capture the Black bishop. After defending the bishop, for example with **17...Qd6**, White would still proceed with **18. g4**, advancing their pawn advantage and achieving an excellent position compared to Black).*

How Not to Play with the Queen.

The following game will serve as an example to show that it is a mistake to place the queen in front of your own minor pieces.

The game began with **1. d4 d5 2. Nf3 Nf6 3. c4 c6 4. Qc2 dxc4 5. Qxc4 Bf5 6. g3 e6 7. Nc3**

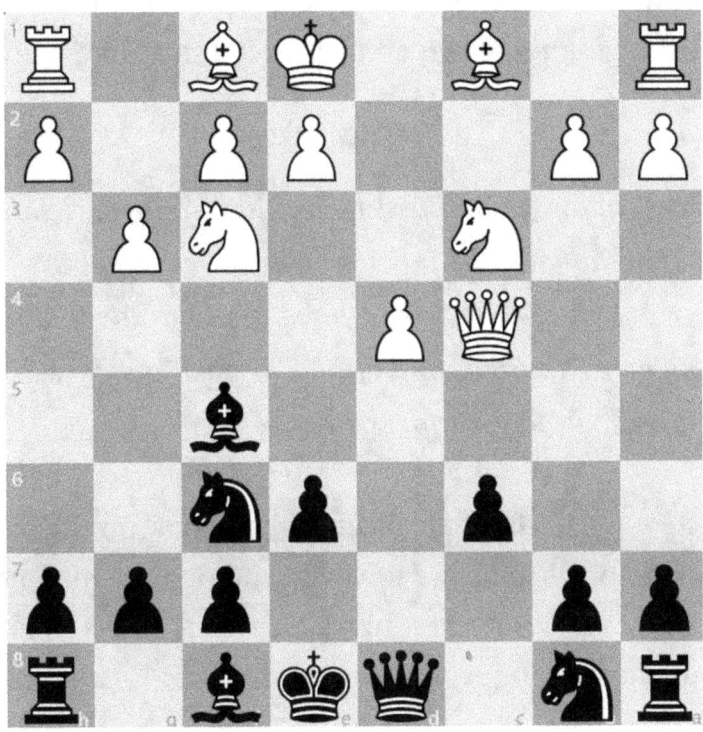

White's last move has a problem, as the white queen will be poorly placed in front of the knight, making it more exposed to attacks from the black pieces and making retreat difficult.

As you know, the queen is the most valuable piece, and exposing it early in the game is dangerous.

Generally, the queen is developed later, and bringing it out so early and in front of your minor pieces is usually too risky.

Black continued their development with **7...Nbd7**, and White followed with **8. Bg2**, developing their bishop and preparing to castle. Now Black played **8...Be7**, also preparing to castle. White castled with **9. O-O**, and Black did the same with **9...O-O**.

*(The correct move for White now would have been **10. Re1** with the idea of advancing the pawn to e4, thus occupying the center with their pawns. White would have a slight advantage due to their space advantage, but Black's position is very solid, making it difficult to exert pressure).*

But White played **10. Rd1**, placing their rook on the d-file and targeting the black queen. However, although it may not seem so, this move by White is a mistake, and now it's your turn to figure out why.

How would you continue with the black pieces?

Take a few minutes and when you have a plan, look at the next page.

Black continued with the move **10...Bc2!**, and White resigned. Since Black is attacking the white rook and White cannot avoid the loss of material, for example, if White continues with **11. Rd2** to move the rook out of the threat and attack the bishop on c2, Black would continue with **11...Nb6**, attacking the white queen. Now it becomes clear how poorly placed the white queen was. The black bishop on c2 controls the d3 and b3 squares, so the white queen would have no squares to retreat to, and Black would be able to capture it, gaining a winning advantage.

Pawn moves that weaken the castling position.

Now we will study the weakening pawn moves in the castling position.

The strength of pawns is at its highest when they are on their original squares. However, once the pawns advance, their defensive power decreases because weaknesses arise that your opponent can exploit to attack.

Let's look at the following position.

For example, here Black played **1...h6,** a move often made to give the king an escape square and to avoid a back-rank checkmate.

However, special attention must be paid to pawn moves, as I've mentioned in previous sections of this book. Pawns cannot move backward, and each time a pawn advances, it inevitably creates weaknesses in the squares adjacent to it.

Before making any pawn move, it is important to apply "the triple rule".

The triple rule states that before making any pawn move, you should ask yourself if that pawn move will facilitate your opponent's attack through one of the following points:

1.- Weakening squares.
2.- Inducing sacrifices.
3.- Creating deployment points, meaning you need to consider if any square will become weak.

In the example illustrated on the previous page, the g6 square has become weakened by the advance of the h6 pawn. Previously, it was protected by both the f7 and h7 pawns, but now it is only defended by the f7 pawn.

Now, let's look at the same structure for Black but with more pieces on both sides.

In this position, you can see that Black has the same structure as in the previous example, but now there are more pieces on the board. Black's last move was **1...h6**, and the g6 square is less protected because it is now defended only by the f7 pawn. However, this pawn is now pinned by the White bishop on c4.

So now White could play **2. Qg6**, threatening checkmate on g7. Black cannot capture the queen because their pawn is pinned, so White would win.

The only way to prevent the mate on g7 is for Black to sacrifice their queen with **2...Qg5**, and White would capture it and win easily anyway.

Another position that can serve as an example is the following:

In this example, you can see how Black advanced their pawn to g6, which weakened the f6 and h6 squares. This allowed White's pieces to invade and exploit these weaknesses, eventually enabling a checkmate with the queen on g7.

Now, the next example is as follows:

In this position, White played **1.g3**, weakening the f3 and h3 squares, which Black then exploited by occupying them with their pieces, leading to checkmate.

Let's continue with the next example:

One of the points mentioned earlier is that pawn advances can provoke your opponent into sacrificing a piece

For example, in this position, Black played **1...h6**, and White can take advantage of this pawn advance with **2. Bxh6**, and after **2...gxh6**, White would respond with **3. Qxh6**. In the next move, they would deliver a check to the black king on the open "g" file, and the black king lacks protection from its pawns because White has eliminated them thanks to the sacrifice made with their bishop on h6.

And if Black attempts to provide an escape square for their king with **3...f6**, White would still play **4. Rhg1+**, and after **4...Kf7**, White would deliver checkmate with **5. Qg6#**.

Alternatively, if instead of advancing their pawn to provide an escape square for the black king, Black plays **3...Ne5**, intending to cover with their knight when White checks with the rook, then White could play **4. Nxe5** first. After Black recaptures with **4...dxe5**, White would deliver checkmate with **5. Rhg1#**.

Let's continue with the next example:

In this new position, Black advanced their pawn to the g6 square, also creating a point where there could be a sacrifice. After White plays **1. Nxg6**, Black recaptures with **1...fxg6**, then White sacrifices again with **2. Bxg6**. If now Black plays **2...hxg6**, White continues with **3. Qxh6+**, and after **3...Qg7**, which is the only move, White delivers checkmate with
4. Qxg7#.

*(And if after **2. Bxg6**, Black plays **2...Qg7**, White plays*
***3. Bxh7+**, and after **3...Kf7**, White continues with **4. Qxg7+**, Black moves their king with **4...Ke8**, then White delivers check with **5. Bg6+**, Black tries to hide their king with **5...Kd8**, and after **6. Rh7**, White threatens checkmate with the queen on d7. If the knight is moved, for example, with **6...Nc5**, then White captures the rook with **7. Qxf8#**, resulting in checkmate).*

Let's continue with the next example position:

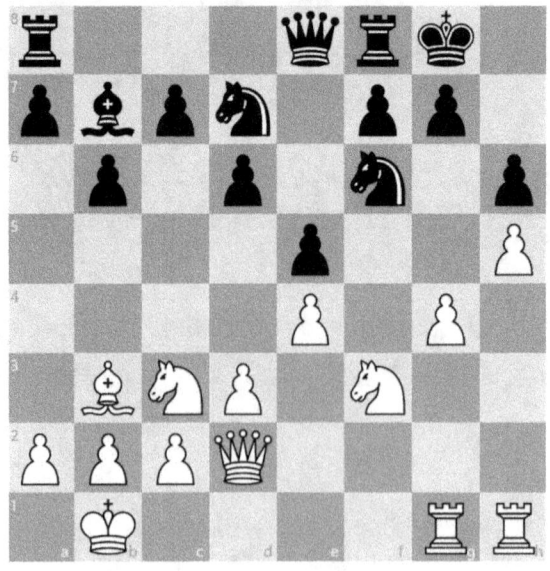

Another consequence of pawn advances in castling is that they create deployment points that your opponent can exploit to open lines and attack the king. For example, in this position, Black has advanced their pawn to h6, creating a deployment point on the g5 square. When White plays **1. g5**, there will be a pawn exchange, thereby opening the column.

After **1...hxg5**, White would continue with **2. Qxg5**, threatening checkmate on g7, which would be inevitable.
If Black chooses **2...g6**, White could then proceed
with **3. Qxg6+** because the f7 pawn cannot capture as it is pinned. After **3...Kh8**, White would deliver checkmate with
4. Qg7#.

Let's continue with another position:

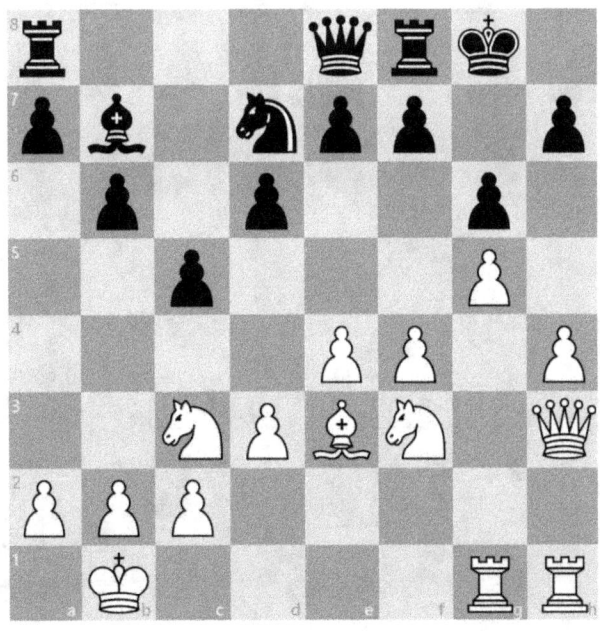

In this position, we can observe that Black has advanced their pawn to g6, creating deployment points on the h5 and f5 squares that White can exploit to attack.

For example, White continued with **1. h5,** in the next move, they will capture the pawn and open the "h" file for an attack against the black king.

In this case, the deployment point on h5 has allowed the opening of the file because the g6 pawn was blocked and could not advance.

Now, let's analyze a similar position where Black's pawns are in their initial squares with castling.

We can observe that Black's pawns are in their perfect formation, and if White were to attempt an attack with the move **1. h6**, thinking of capturing the pawn to open the "h" file as in the previous example and then deliver checkmate on h7, Black could easily play **1...g6** and maintain the closed position.

This is what the triple rule consists of.

So, every time you are about to move a pawn, you will be able to analyze whether it benefits you or not.

Since you will be more attentive to the squares that you will likely leave weak.

With these examples, you are now more aware of pawn advances and can take advantage when your opponents leave vulnerable squares.

Types of center.

As you may well know, the center is very important in chess, and from the early moves, one should try to control the central squares.

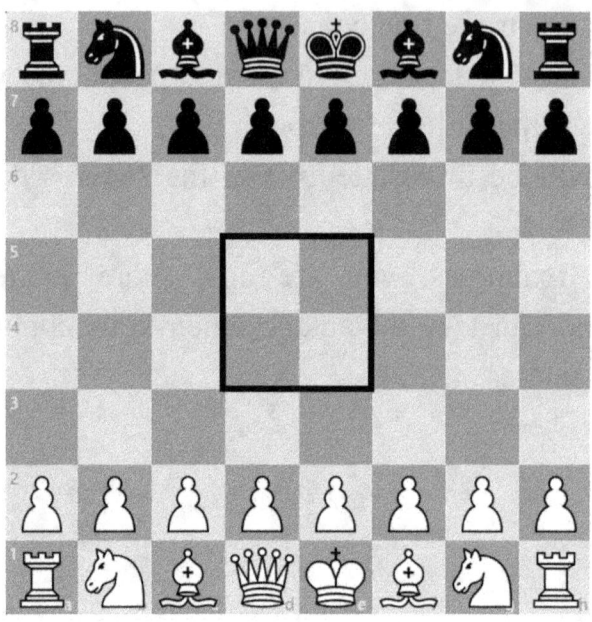

In this image, you can see the four central squares, which are very important, and from the beginning of the games, both players will have to try to control them.

It is generally recommended to advance with the central pawns, for example, with the move e4, and if possible, also play d4, thus occupying the center with pawns and dominating several squares in the center.

However, both players will try to fight for the center, and after a few moves, the structure will be defined, and many variations of structures can occur.

Now, let's analyze what types of structures can occur and how to play with each one.

Structure #1
The classical center.

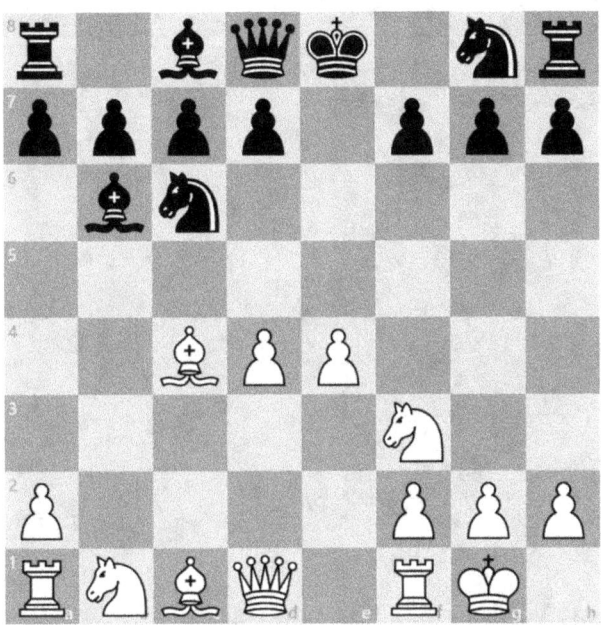

This type of center occurs when one of the two players has placed two pawns in the center.

In this position, it's the white player who has placed two pawns in the center, and generally, establishing a classical center in this way tends to confer an advantage as it allows good control over the central squares and provides more space to the player who has this type of center.

This center can be very dangerous because if the opponent does not prevent the pawns from advancing, they can advance, creating many dangers, and forcing the opponent's pieces to retreat. It could be the beginning of an attack that could be decisive.

Structure #2
The small center.

In this position, White only has one pawn in the center, and this position is not as strong as the previous one. It's less powerful than the classical center, and Black is not in danger, as the black pieces can easily block or eliminate the e4 pawn.

This center only grants White a little space because they have a pawn on the fourth rank.

With the disappearance of the d-file pawn, this file is now semi-open, and White can place their rooks on this file, creating pressure on it.

Also, the d5 square is an excellent square that could be used, for example, with the white knight, and this knight would serve as a great defender and attacker at the moment, as it is in the center of the board and cannot be immediately expelled by a black pawn because then the knight could go to the b6 square, attacking both the black queen and the rook on a8.

Another characteristic of this position is that White could advance their f4 pawn to help the other pawn control more central squares.

Also, you can observe that White has 4 pawns on the kingside, while Black only has 3.

So White could attack in this area, which means attacking with an extra pawn.

Structure #3
The open center.

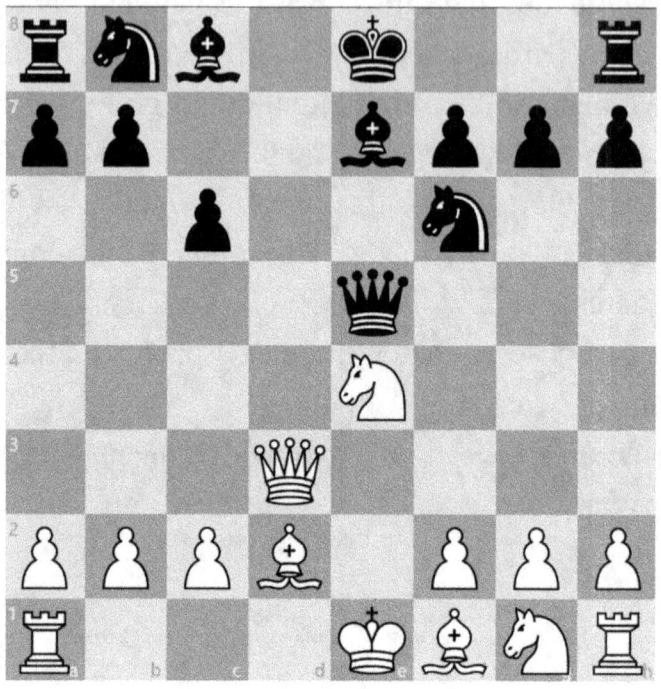

This is the open center because there are no pawns in the center.

The characteristic of this new position is that with no pawns in the center, the central files are open.

With pawn exchanges having occurred, the two central files are free, and in these types of positions, King safety is very important. You need to castle quickly; otherwise, the King could be in danger being on the files that are now open.

A typical plan could be to double rooks on the open files to invade the enemy's area. In this type of position, the attack with pieces is more important than the attack with pawns since it would be very slow with pawns, whereas with pieces, progress would be faster.

Another characteristic to consider in this type of position is that if many pieces are exchanged, then it will lead to an endgame that will be more or less balanced, as the pawn structure is symmetrical, minimizing the chances of creating a passed pawn.
So, a good defensive technique would be to exchange pieces to simplify the position.

Structure #4
The closed center.

We can observe that both White and Black have two pawns in the central files, but they are both blocked, meaning the center has been closed, and in these positions, the pawns cannot advance, and attacking in the center is not possible. So when the center is locked, the game will shift towards the flanks.

In this type of position, King safety is not as important because the King in the center is not as vulnerable since it's blocked and cannot be easily attacked.

Typically, both players will attack on the side where they have more space.

The best advice to determine where to attack is to identify where you have more room to mobilize and maneuver your pieces to attack against the base of the opponent's pawn chain.

For example, in this position, if you observe the pawn chain of the white pieces pointing towards the direction of the kingside, you'll realize that White has more space on this flank; hence, their plan should be directed towards it.

A plan could be to advance their pawn to f4, and then push it to f5 to attack against the black pawn on e6, thus breaking down the black pawn chain and opening up the game on the kingside where they have more space.

Conversely, for the black pieces, their attack should be directed to the queenside, as they have more space in this area of the board. A good plan for Black would be to advance their pawn to c4 to attack against the d4 pawn and thus break the pawn chain of the white pieces.

Structure #5
The fixed center.

In this position, only one central pawn remains for each player, but both are blocked, therefore they cannot move, hence the center is fixed.

In this position, it's not necessary to shift strategies towards the flanks, as there is an open file in the center.

The characteristics of this type of center are that there is an open file, so it's important to occupy it with the rooks, and also the central pawn, for example for White, controls two squares (d5 and f5), and these squares are good for placing pieces.

In conclusion, one should occupy the open file with the rooks and try to bring pieces towards the squares controlled by the central pawn.

Structure #6
The dynamic center.

In this type of position, the center is not yet defined, which is why it's called "dynamic," because it can move or different types of centers can arise.

For example, in this position, the center is very tense, as many piece exchanges or pawn advances can occur.

When the pawns are in contact and exchanges can occur in the center, then it can be said that there is tension in the center.

In this position, if White decides to play **1. d5**, then it would transpose to a locked center.

Alternatively, instead of advancing their pawn, if they play **1. dxe5**, and Black recaptures with **1...dxe5**, it would transpose to a fixed center, both players with a pawn in the center but blocked, and there is also an open file.

Alternatively, White could make another move, maintaining the tension and keeping the center dynamic.

With this, you are aware of the most common types of centers that can arise in your chess games. Now that you know where to direct your attack when facing them, you can start applying it in your future games.

Evaluate the position.

Carefully analyze the following position and consider the weak points and areas where you can attack.

Take 5 minutes.

The true strength of a chess player is measured by their accuracy when evaluating positions.

A thorough evaluation of any position will lead you to make better decisions, and therefore, to make the best moves.

Nothing ruins a good position faster than misjudging it, thinking you have an advantage or that there is no tactic involved.

For example, in the previous position you analyzed, the black pieces lost some tempo in the opening and had to move their king to the square, only to return it to e8 later.

Therefore, they cannot castle, and the white pieces thought they had an advantage and tried to exploit it.

So, the white pieces played **1. Bc2,** with the idea of bringing their queen to the d3 square, and then invading the h7 square, attempting to gain an advantage.

But the black pieces continued with **1...Qc3**, threatening to capture the c4 pawn.

(And now the best move for the white pieces would have been to reach an equal endgame with **2. Qd3***, and after* **2...Qxd3** *and* **3. Bxd3***, the position would be balanced).*

However, in the game, the white pieces did not want to reach that position because they thought they had an advantage, and decided to continue with the move **2. Bb3** to defend the c4 pawn.

The black pieces continued with **2...Kf8**, removing their king from the open file, and the white pieces played **3. Rc1**, attacking the black queen to drive it away, and the black pieces retreated it with **3...Qf6.**

The white pieces continued with **4. Bc2**, again with the idea of bringing their queen to the d3 square and putting pressure on the h7 square, but the black pieces continued improving their pieces with **4...Rae8**, doubling their rooks on the open file.

The white pieces played **5. Qd3**, and the black pieces continued with **5...Bg4**. The black pieces continue calmly improving their pieces in good squares, and the position is actually balanced, and the white pieces do not have an advantage as they think.

*(For example, if the white pieces were to continue now with their plan with **6. Qh7**, trying to invade with their queen, the black pieces would simply play **6...g5**, and the black queen controls the h8 square, also defends the h6 pawn, and they have everything perfectly under control. And then they could improve their dark-squared bishop to the b4 square, attacking the white rook on e1, and the black pieces would indeed have an advantage).*

That's why the white pieces continued with **6. Bd2**, trying to control the b4 square, but they did not realize that the black pieces have a tactical resource.

The move made by the black pieces was **6...Re2,** and they are winning.

The black pieces are threatening to capture the f2 pawn, which would lead to checkmate.

*(and it would not be possible for the white pieces to play 7. **Rf1** to defend the pawn because the black pieces would continue with 7...**Rxd2**, and after 8. **Qxd2**, the black pieces would play 8...**Qf3+**, and after 9. **Kg1**, the black pieces would continue with 9...**Bh3**, and in the next move, it would be checkmate).*

In the game, the white pieces continued with **7. Rxe2**, removing one black rook, and the black pieces recaptured with the other rook with **7...Rxe2**, resuming the threat against the f2 pawn.

*(And it would not be possible for the white pieces to defend it with 8. **Be1**, because the black pieces would then follow with 8...**Bc5**, and the black pieces would attack the f2 pawn with three pieces and would be winning.*

*And if after the move 7...**Rxe2**, the white pieces play 8. **Be3**, the black pieces would play 8...**Qf3+**, and after 9. **Kg1**, the black pieces would play 9...**Rxc2**, removing the white bishop and preventing it from supporting the white queen to go to the e4 square from where it would defend the mate on g2, and after the white pieces recapture with 10. **Qxc2**, then the black pieces would continue with 10...**Bh3**, and in the next move, they would deliver checkmate on the g2 square).*

That's why in the game, after the move **7...Rxe2,** the white pieces played **8. Rf1**, trying to defend the f2 square, but the black pieces played **8...Rxd2**, and the white pieces resigned.

Since after **9. Qxd2,** the black pieces would proceed with **9...Qf3+**, and after **10. Kg1**, the black pieces would continue with **10...Bh3**, and the white pieces are completely lost.

The only way to stop the checkmate is by playing **11. Be4,** and the black pieces would then play **11...Qxe4**, and the white pieces would have to play **12. f3** to avoid checkmate.

Then, the black pieces could continue, for example, with **12...Qf5**, threatening to capture the rook on f1 with their bishop. It wouldn't be possible for the white pieces to retreat their rook with **13. Re1** because the black pieces would follow with **13...Bc5+**, and the black pieces win. After the white pieces move their king with **14. Kh1**, the black pieces would play **14...Qxf3+**, and after the white pieces cover the check with **15. Qg2**, the black pieces mate with **15...Qxg2#**.

And if after the move **12...Qf5**, the white pieces play **13. Rf2** to remove their rook from the threat of the bishop on h3, the black pieces would simply continue with **13...Qb1+**, and the black pieces would win, as any piece covering the check is useless. For instance, after the white pieces cover the check with **14. Qe1**, the black pieces would play **14...Qxe1+**, and after the white pieces cover again with **15. Rf1**, the black pieces mate with **15...Qxf1#**.

So, with this example, you can notice how easy it is to make poor decisions in positions where we think we have an advantage, when in reality, the position is balanced or even slightly disadvantageous.

The importance of analyzing all your positions at any part of the game is vital to increase your strength in your game; the right decisions come when the position is properly analyzed, not only in attack but also in defense, to escape from a disadvantageous position and achieve equilibrium.

Endgames

Let's start with rook endgames.
So, analyze this position and find the way in which the black pieces can force a draw.

In this position, White plays **1. e5,** and Black responds with **1...Rb6**.
The tactic that Black should follow is to keep their rook on the sixth rank until the pawn of White advances.

White proceeds with **2. Ra7,** and Black continues waiting on the sixth rank with **2...Rc6,** as long as the pawn of White doesn't advance.

If finally the pawn of White advances with **3. e6**, then Black must give checks from behind to the White king. Therefore, Black would play **3...Rc1,** White would continue with **4. Kf6**, threatening mate, but Black would play **4...Rf1+**, and since the pawn advanced to the sixth rank, the White king can't find anywhere to hide from the checks.

For example, **5. Ke5**, Black would follow up with **5...Re1+**, **6. Kf5, Rf1+, 7. Kg4.** When the king attempts to approach the Black rook, then the rook would attack the pawn with **7...Re1**, and White must bring their king back to defend the pawn.
It would continue with **8. Kf6**, but Black would play **8...Rf1+**, and so on until a draw is reached.

Queen vs Advanced Rook Pawn.

In this position, you will learn an endgame where the opponent has a pawn on the seventh rank, about to promote. However, the best possible outcome is a draw.

The technique you can use to win such positions is called "The Staircase."

In this case, since it is a rook pawn that is about to promote, this technique cannot be applied.

I will show you why it cannot be used.

If the game continues as follows:

1. **Qg7+, Kf2**
2. **Qh6, Kg2**
3. **Qg5+, Kf2**
4. **Qh4+, Kg2**
5. **Qg4+, Kf2**
6. **Qh3, Kg1**
7. **Qg3+, Kh1**

You can observe that the Staircase method has been used; however, the black king is forced to stand in front of its pawn, blocking its advancement.

This would be the moment in the Staircase technique to bring the king closer, but the problem is that with the advanced pawn being a rook pawn, the black king would end up stalemated.

Therefore, this technique cannot be used for this type of position.

However, if in the same position the white king were closer, the situation could end differently.

Let's analyze it then:

In fact, as long as the white king is within a certain imaginary square range, the position will be winning for White.

To understand this better, you can observe the winning zone in the following image:

So, having said this, let's see how White would win: They would start by applying the Staircase technique:

1. **Qg7+, Kf2**
2. **Qh6, Kg2**
3. **Qg5+, Kf2**
4. **Qh4+, Kg2**
5. **Qg4+, Kf2**
6. **Qh3, Kg1**
7. **Qg3+, Kh1**

And now, if White makes an imprecise move, the black king would be stalemated again, so careful consideration is required for how to continue.

Therefore, it should be continued with:

8. Ke1+ , Kg2
9. Kf4

And now, White can bring their king closer.

Black will promote their pawn to a queen, but then White will continue with **9...Qe2+**, and if the king moves to **10. Kh3**, Black will play **10...Qg4+** and after **11. Kh2**, Black checkmates with **11...Qg3#**.

And if instead of moving to h3, White plays **10. Kg1**, Black will play **10...Kg3**, and checkmate is inevitable.

Rook and Bishop vs. Rook.

The rook and bishop vs. rook endgame is a draw if played correctly.

A straightforward defense system is called the Cochrane Defense, which involves the following:

In this position, White is about to bring their king and bishop closer to attempt a checkmate.

If it were Black's turn to move, the technique involves keeping the rook on the column where both kings are located.

For example, now Black would make a waiting move like:

1... Rd2, the rook stays on the file where both kings are, and when the king moves away, the technique involves the black king moving to the opposite side. That is, if White moves their king to **2. Ke5**, Black would play **2...Kc8**, and if instead of moving their king to e5, White plays **2. Kc5**, Black would move theirs to **2...Ke8**, always to the opposite side.

Now, after White moves their bishop to **3. Be5**, Black should move their rook to the second rank with **3...Rd7**; White's rook is under attack and they would have to move it, for example, with **4. Rh8+**. The black king moves out of check with **4...Kf7**, and it's been liberated.

Let's return to the initial position but now with the difference that it's White's turn to move.

If it were White's turn to move, it would be similar to what we've seen before, for example:

After **1. Ke5**, then Black would continue with **1...Kc8**, always moving to the opposite side.

And after White moves their bishop, let's assume to **2. Bc5**, Black brings their rook to the second rank with **2...Rd7**, and in the next few moves, the black king will be liberated.

If White checks with **3. Rh8+**, Black frees their king with **3...Kb7**.

And if White tries to complicate the situation and instead of checking, plays **3. Be7**, Black would still play **3...Kb7**, and after White plays **4. Ke6**, attacking the black rook, Black defends it with **4...Kc6**, and White plays **5. Rh1**, threatening to check the black king on c1 and then capture the black rook with the king, so Black removes their rook with **5...Rd2**, and the game continues. **6. Rc1+, Kb5 7. Bd6** (trying to shield from the rook's checks with the bishop) but it would still continue **7...Re2+**, if White covers the check with the bishop, with **8. Be5**, then Black would play **8...Rd2**, preventing the white king from advancing.

Now, if instead of covering the check with the bishop, White plays **8. Kd7**, the game would continue as follows:

8...Re4
9. Rc5+, Ka4
10. Kc6, Kb3

11. Kd5, Re8
12. Rb5+, Kc2
13. Bc5, Kd3
14. Rb3+, Ke2
15. Bd4

And now it seems that White is succeeding in trapping the black king on this part of the board, but after Black plays **15...Rd8+**, it would continue with **16. Ke4, 16...Re8+ 17. Be5**, and after **17...Ke1** and **18. Rb2**, we have the same position as at the beginning but on the opposite side of the board.

Black would calmly wait with **18...Re7**, staying on the file where both kings are located, and they continue with this technique. When the white king moves to one side, the black king must move to the opposite side. And the game ends in a draw.

Endgames with knights.

The winning method is to remove the opposing pieces from controlling the promotion square of the b7 pawn.

So, White plays **1. Ng6**, threatening to bring their knight to the f8 square to divert the black knight, and they are also threatening to go to e5, checking the black king and attacking the black knight, which would also divert it from defending the b8 square.

That's why Black plays **1...Kd5**, and White continues with **2. Nf8**, attacking the black knight and attempting to divert it. Now, the only correct move for Black would be **2...Ne5**.

(Of course, it would be a huge mistake to capture the white knight with **3. Nxf8**, because then White would promote to a Queen and win the game easily).

So, after Black played **2...Ne5**, White cannot promote now, because if they do with **3. b8=Q**, then Black achieves a draw with **3...Nc6+** and would later capture the queen.

White can now win in several ways; for example, one of the moves could be **3. Kb6**, followed by Black's **3...Nc6**, which is the only move at the moment preventing pawn promotion. Then, White continues with **4. Nd7** and Black with **4...Kd6**; the black king must continue defending the knight. Now, White can proceed with **5. Ne5!** An excellent diversion move!

(If the black king captures the white knight with **5...Kxe5**, then White captures the black knight with **6. Kxc6**, and the promotion of the white pawn is unstoppable.
And if instead of capturing the white knight with the king, Black captures it with their knight with **5...Nxe5**, then White promotes to a queen with **6. b8=Q** and wins).

Therefore, the only move for Black would be **5...Nb8** because White was threatening to exchange the knights.

White continues with **6. Ka7**, threatening to capture the black knight, which cannot move. Therefore, Black must defend it with the king by playing **6...Kc7**, and White would then proceed with **7. Nc4**.

*(If now Black were to play **7...Nd7**, White would play **8. Nb6**, and again, the black knight cannot capture it because then White would promote. Therefore, the only move would be to retreat the knight with **8...Nb8**, but after White plays **9. Nd5+**, the black king has to move away from defending the knight, and in the next move, White would capture it and win).*

So after the move **7. Nc4**, Black played **7...Nc6+**, and White moves out of check with **8. Ka8**.

*(If Black were to move their king, for example, to square **8...Kd7**, then White would play **9. Ne5+**, and they would manage to exchange the knights and promote their pawn, winning).*

Therefore, after **8. Ka8**, the only move for Black is **8...Nb8**, and White would continue with **9. Nb6**, and of course, Black cannot capture the knight with their king, because then White would capture the black knight and win.
So Black has only two possible moves, to bring their knight to either a6 or c6. Let's see the variations:

Let's start with c6.

If Black plays **9...Nc6**, then White continues with **10. Nd5+**, and after the black king moves **10...Kd7**, White would play **11. Nb4** and ultimately divert the black knight and win.

Now let's analyze if it goes to square a6.

If Black plays **9...Na6**, White would play **10. Nd5+**, and if the king moves, for example, to **10...Kd6**, White would continue with **11. Nb4**, and White would also manage to divert the black knight and win.

Now, let's go back several moves to analyze other possibilities. For example, after White's first move, if instead of playing **1...Kd5**, Black plays **1...Kc5**.

White would continue with similar moves.
They would play **2. Nf8,** attacking the black knight. Black would respond with **2...Ne5**, momentarily preventing White from promoting because if they do, the black knight would check on c6, capturing the queen and achieving a draw.
So White plays **3. Ka8,** and Black would have to play **3...Nc6**, the only move to prevent the pawn from promoting, and White would continue with **4. Nd7+** and Black with **4...Kd6**.

(Now it would be a mistake for White to play **5. Nb8**, *because Black would play* **5...Nb4**, *and the white knight would have no squares. So White would play* **6. Ka7**, *and Black would continue with* **6...Kc7**, *achieving a draw).*

That's why White continued with **5. Nb6**.

(If now Black played **5...Kc7**, *White would continue with* **6. Nd5+**, *and after* **6...Kd6**, *White would play* **7. Nb4**, *and White would win as seen earlier.*

And if after **5. Nb6**, *Black plays* **5...Kc5**, *White would play* **6. Nc8**, *and in the next move, they would bring their knight to e7, diverting the black knight and promoting).*

This endgame is important to know and study because you must perfectly understand the technique of diverting pieces that prevent promotion in endgames.

Bishop vs. Knight Endgame.

In this position, both sides are evenly matched, except that White has a knight and Black has a bishop.

Apparently, the black bishop has enough ability to control the white pawn and prevent it from promoting.
However, let's study this position to show you how it's possible for White to gain an advantage from it.

White plays **1. d6**, advancing their pawn, and if Black does nothing, it can advance to d7, leading to an inevitable promotion.

*(It wouldn't be possible to prevent this advance with **1...Be6**, because White would play **2. Nc6+** and then capture the bishop, easily winning).*

That's why the only move for Black is **1...Be8**, controlling the advance of the pawn.
White continues with **2. Nf6**, attacking the black bishop.

*(It wouldn't be a good move to retreat the bishop with **2...Bb5**, because then White would play **3. Kc5** and Black is in zugzwang, as between the white king and the knight, they control all squares on the bishop's diagonal. The black king couldn't move, as it would stop protecting the bishop, and White could capture it, resulting in a win).*

So, after White's move **2. Nf6**, Black has to play **2...Bc6**, and after White plays **3. Kc5**, attacking the bishop, Black can play **3...Bb5**, unlike the previous variation. So, White continues with **4. Kb6**.

Now, Black cannot move their bishop because there are no available squares to continue protecting the pawn's advance except for the square it occupies. Therefore, the only move is to play **4...Kb4**, to continue defending their bishop.

And now White continued with **5. Nd5+.**

*(It wouldn't be possible to get out of check with **5...Ka4** because White would continue with **6. Nc3+** and in the next move, they would capture the bishop, and the pawn would promote without issues)*

After the check, Black played **5...Kc4**, the black king must continue maintaining the defense of its bishop, and White continued with **6. Ne3+**. Black retreats with **6... Kb4**, but White gives another check with **7. Nc2+**.

*(Now, it wouldn't be possible for Black to play **7...Kc4**, because White would then play **8. Na3+**, capturing the bishop in the next move and winning).*

So, Black played **7...Ka4**, and White now plays **8. Kc7**.

Now, in this position, the black king cannot move because if it moves to the only available square, b3, White can check with their knight on d4 and would capture the black bishop.

So, Black has to move the bishop, but obviously not to any square, it must maintain control over the pawn's advance. That's why Black continued with **8...Be8**.

But now White played **9. Kd8**.

*(And now the black bishop cannot stay on that diagonal because if it goes to square **9...Bc6**, White would play **10. Ke7**, and the black king still cannot move because both **10...Kb5** or **10...Kb3** would allow the knight to check with a double attack, and the black bishop would be captured, resulting in a win for White.*

*So, Black can only play **9...Bb5**, and after **10. Nd4**, White is already winning because between the knight and the king, they control all squares on the bishop's diagonal, and in the next move, they capture it, promoting the pawn, and the game is won for White).*

So, Black continued with **9...Bh5**, trying to switch diagonals with the bishop, but now White plays **10. Ne3**, preventing the black bishop from going to g4, so Black must continue with **10...Ne2** to return to the initial diagonal, and after **11. Ke7**, Black would play **11...Bb5**, and now White would play

12. Nc2, and with this maneuver, White has returned to a position similar to the variation considered on the previous page but with White on e7.

The only move that resists here for Black is **12...Bc6**, but White would continue with **13. Nd4** and win.

The black bishop is practically trapped, as it no longer has squares to be safe and continue protecting the advance of the white pawn. Therefore, it will be able to advance, promote, and win the game for White.

Tactic, tactic, tactic and tactic.

Chess tactics are combinations of moves that allow the player to win material, checkmate, or significantly improve their position on the board. These maneuvers typically involve short and concise sequences of moves that exploit weaknesses in the opponent's position.

Main Chess Tactics

1. Pin

A pin occurs when a lower-value piece is directly attacked and cannot move without exposing a higher-value piece behind it. This type of tactic is especially effective with rooks, queens, and bishops.

2. Fork.

A fork is a move in which a single piece, usually a knight, simultaneously attacks two or more enemy pieces. This move forces the opponent to choose which of their pieces to save.

3. Discovered Attack.

A discovered attack occurs when a piece moves, revealing an attack by another piece behind it. This type of tactic can be devastating if the discovered piece is a queen or rook.

4. X-Ray.

Also known as an indirect attack, an x-ray occurs when one piece attacks another through a third piece. Rooks, queens, and bishops are particularly effective at executing this type of attack.

5. Deflection.

Deflection involves luring an enemy piece out of its ideal position, allowing for an additional tactic such as a fork or pin. This is typically achieved by offering a piece as bait.

6. Overloading.

An overloaded piece is one that has to defend multiple threats simultaneously. By forcing this piece to choose between two defenses, material can be won or a better position achieved.

Importance of Tactics in Chess.

Tactics are essential for success in chess because they can drastically alter the course of a game. Players who master tactics can capitalize on their opponents' mistakes, win material, and create opportunities for checkmate. Additionally, tactical knowledge helps in avoiding traps and defending against enemy attacks.

Place the following setups on your physical board and begin recording the correct continuations, and at the end, you'll find the answers.

(Not all exercises have 4 moves).

1)

Black to move.

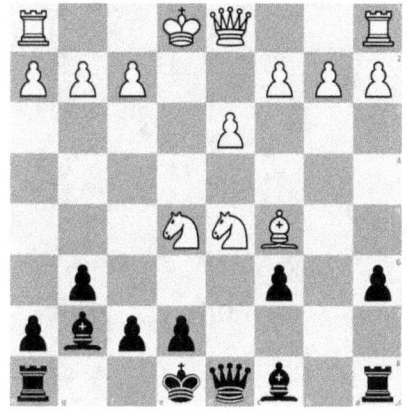

	White	Black
1.	••• _____	_____
2.	_____	_____
3.	_____	_____
4.	_____	_____

2)

Black to move.

	White	Black
1.	••• _____	_____
2.	_____	_____
3.	_____	_____
4.	_____	_____

3)

	White	Black
1.	_____	_____
2.	_____	_____
3.	_____	_____
4.	_____	_____

4)

	White	Black
1.	_____	_____
2.	_____	_____
3.	_____	_____
4.	_____	_____

5)

	White	Black
1.	_____	_____
2.	_____	_____
3.	_____	_____
4.	_____	_____

Black to move.

6)

	White	Black
1.	...	
2.		
3.		
4.		

White to move.

7)

	White	Black
1.		
2.		
3.		
4.		

White to move.

8)

	White	Black
1.		
2.		
3.		
4.		

Black to move.

9)

	White	Black
1.	...	
2.		
3.		
4.		

Black to move.

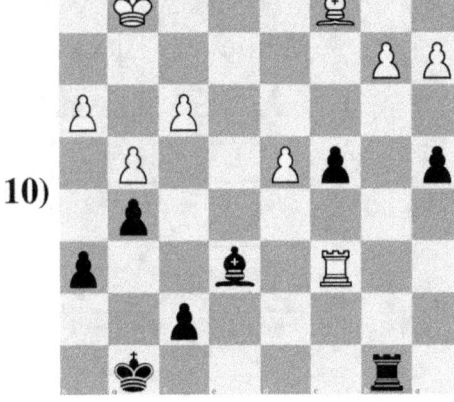

10)

	White	Black
1.	...	
2.		
3.		
4.		

White to move.

11)

	White	Black
1.		
2.		
3.		
4.		

White to move.

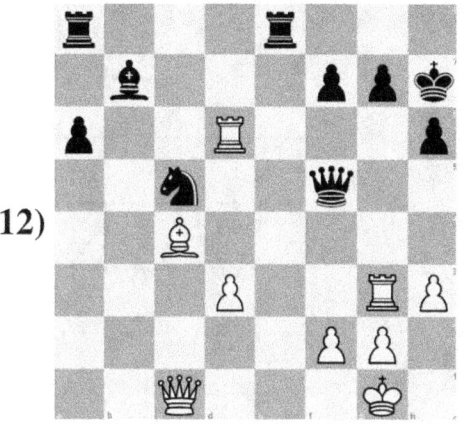

	White	Black
1.	_____	_____
2.	_____	_____
3.	_____	_____
4.	_____	_____

Black to move.

	White	Black
1.	... _____	_____
2.	_____	_____
3.	_____	_____
4.	_____	_____

White to move.

	White	Black
1.	_____	_____
2.	_____	_____
3.	_____	_____
4.	_____	_____

White to move.

	White	Black
1.	_____	_____
2.	_____	_____
3.	_____	_____
4.	_____	_____

15)

White to move.

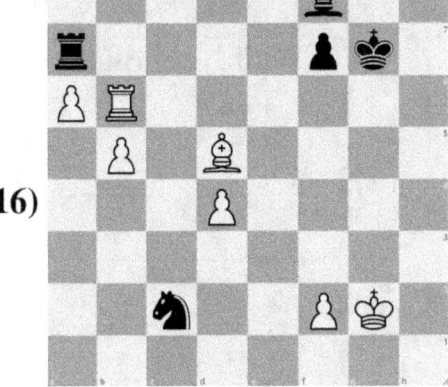

	White	Black
1.	_____	_____
2.	_____	_____
3.	_____	_____
4.	_____	_____

16)

White to move.

	White	Black
1.	_____	_____
2.	_____	_____
3.	_____	_____
4.	_____	_____

17)

Black to move.

18)

	White	Black
1.	...	_____
2.	_____	_____
3.	_____	_____
4.	_____	_____

White to move.

19)

	White	Black
1.	_____	_____
2.	_____	_____
3.	_____	_____
4.	_____	_____

White to move.

20)

	White	Black
1.	_____	_____
2.	_____	_____
3.	_____	_____
4.	_____	_____

SOLUTIONS

01) 1...Qxd5, 2. d4, Bxe5

02) 1...Bxf3 2. Qxe2, Bxe2

03) 1. Bxb3, Bxb3 2. c4, Bxc4 3. Qxc4

04) 1. Qxg8+, Kxg8 2. Rf8#

05) 1. Rf4, Qxf4 2. exf4

06) 1...c6+ 2. Kc5, b6+ 3. Kc4, b5+, 4. Kc5, Nb3#

07) 1. Qxe6+, Kd8 2. Rxf8+, Bxf8 3. Qe8#

08) 1. Qxe8+, Rxe8, 2. Rxc5

09) 1...Rxa2+ 2. Kxa2, Qb2#

10) 1...a3 2. bxa3, Rb1

SOLUTIONS

11) 1. Qxe6+, 2. Rf7, g6

12) 1. Rxg7+, Kxg7 2. Qxh6+, Kg8 3. Rg6+, Qxg6 4. Qxg6+

13) 1...f5+ 2. Kxf5, Kxd3

14) 1. Qf6, Qe1+ 2. Bf1, Rxg2+ 3. Kxg2

15) 1. Qxh6, gxh6 2. Bf5

16) 1. Rb7, Rxb7 2. Bxb7

17) 1. Ng8+, Kg6 2. f5#

18) 1... Qh2, Rxe6 2. Rxe6

19) 1. Qxe4, Qxa1 2. Qxe8+, Rxe8 3. Rxa1

20) 1. Bb5+, Nc6 2. Qxd4

We have concluded the basic strategies you needed to know in your journey to becoming a better chess player.

Please consider leaving your opinion and review of this book, so we can take it into account for the following books in this series.

If the strategies presented here were useful, stay tuned for the next parts of **"How To Play and Win at Chess Like a Master."**

With love for my readers

www.ingramcontent.com/pod-product-compliance
Lightning Source LLC
Chambersburg PA
CBHW071511220526
45472CB00003B/979